FEEDPADS
for Grazing Dairy Cows

JOHN MORAN AND SCOTT MCDONALD

CSIRO

PUBLISHING

National Library of Australia Cataloguing-in-Publication entry

Moran, John, 1945–

Feedpads for grazing dairy cows/by John Moran and Scott McDonald.

9780643097681 (pbk.)

Includes bibliographical references and index.

Dairy cattle – Feeding and feeds – Australia.
Dairy farming – Australia.
Cattle – Feed utilization efficiency – Australia.

McDonald, Scott.

636.21420994

Published by
CSIRO PUBLISHING
36 Gardiner Road, Clayton VIC 3168
Private Bag 10, Clayton South VIC 3169
Australia

Telephone: [+613] 9545 8555
Local call: 1300 788 000 (Australia only)
Fax: +61 3 9662 7555
Email: csiropublishing@csiro.au
Web site: www.publishing.csiro.au

All photographs are by the authors.

Set in Adobe Minion Pro 11/13.5 and Adobe Helvetica Neue
Edited by Janet Walker
Cover and text design by James Kelly
Typeset by Desktop Concepts Pty Ltd, Melbourne
Printed by Ingram Lightning Source

CSIRO PUBLISHING publishes and distributes scientific, technical and health science books and journals from Australia to a worldwide audience and conducts these activities autonomously from the research activities of the Commonwealth Scientific and Industrial Research Organisation (CSIRO).

The views expressed in this publication are those of the author(s) and do not necessarily represent those of, and should not be attributed to, the publisher or CSIRO.

Foreword

Dairy systems in Australia are facing ever-increasing feed and operational costs. In south-east Australia these costs have been driven by the impact of continuing drought reducing the availability of irrigation water, and the challenge of climate change. In some cases in the southern Murray–Darling Basin, water allocations have declined to such an extent that alternative strategies to feed cows are being examined. Feedpads will play an integral part to these new systems of production and therefore, this book is a welcome and timely contribution to the dairy industry. The dairy industry will adopt two strategies for future efficient production: increased reliance on purchased feeds and a change towards smarter use of irrigation water by developing systems less reliant on traditional perennial rye grass pastures, such as an increase in annual pastures and double-cropping. Both of these strategies will mean that dairy farmers have to increase the efficiency of feeding, reduce feed wastage, maintain animal health and welfare, and manage the impact of increased heat loads resulting from more heat stress events.

Adopting an increased reliance on purchased feeds inevitably leads to an intensification of production as the profitability of these systems is heavily reliant on efficient conversion of feed to milk. The correct design of the feedpad (site assessment, dimensions, use of appropriate construction materials, and management of environmental pollution) is costly, but in the long term, the efficiencies achieved will feed through into the whole farm return on asset. Further, the installation of a feedpad when the system has a greater reliance on purchased feeds allows partial mixed rations to be fed at times of year when water availability is restricted. The alternative strategy of developing a system more reliant on annual pastures and cropping also requires a feedpad system to be efficient. For example in rain-fed dairy systems, the integration of whole crop cereal silage–brassica summer cropping systems has been demonstrated as profitable. These systems also require feedpads to deliver the daily allocation of whole crop cereal silage efficiently to the herd after milking. Wastage of the homegrown feed is therefore reduced, thus increasing the whole farm profitability, as well as the feed conversion efficiency.

Finally, the implication of climate change (increased night-time temperatures as well as reduced rainfall) further supports the development of feedpads in pasture-based dairying. Thermal heat loadings in the dairy cow do not only reflect the daytime temperature and the provision of shade, but also reflect the ability of stock to shed heat during the night. Management of heat load in high performance dairy cows can be improved using a well-constructed feedpad system insofar that it provides the opportunity of shade and, potentially, evaporative cooling. Better management of heat stress inevitably leads to better cow health and welfare. The concern that cows managed on feedpads are 'uncomfortable and unhygienic' is refuted in the text. A well-designed and well-managed feedpad system has been demonstrated by many studies to actually reduce animal health issues and that hygienic milk production is possible.

The authors of this excellent book should be congratulated on its practical approach to the development of feedpads in efficient dairy systems. The information is presented clearly and is supported by many research studies undertaken in Australia and elsewhere.

Julian Hill
Senior Lecturer in Ruminant Nutrition,
Department of Agriculture and Food Systems, Land & Environment, University of Melbourne, Parkville, Victoria

Contents

Acknowledgements

John and Scott would like to thank their colleagues with whom they worked at DPIV Kyabram for their many ideas and contributions to discussions on new feedpad technology. In addition, they would like to thank Steve Little (Manager of the Feed Fibre Future program) and Dan Armstrong (until recently at DPIV Ellinbank in Gippsland) for reviewing the manuscript prior to its publication.

Dr John Moran
Senior Dairy Adviser
Department of Primary Industries Victoria (DPIV)
255 Ferguson Rd, Tatura 3616, Australia
Telephone: +61 (0)358 335 215 (office), +61 (0)418 379 652 (mobile); facsimile:
+61 (0)358 335 299
Email: john.moran@dpi.vic.gov.au

or

24 Wilson St, Kyabram 3620, Vic, Australia
Ph: +61 (0)418 379 652
email: jbm95@hotmail.com

Mr Scott McDonald
Statewide NRM Project manager – Dairy
Farm Services Victoria
Department of Primary Industries Victoria (DPIV)
PO Box 441 Echuca 3564, Australia
Telephone; +61 (0)354 820 440 (office), +61 (0)438 227 779 (mobile); facsimile:
+61 (0)358 335 299
Email: scott.mcdonald@dpi.vic.gov.au

Chemical warning

The registration and directions for use of chemicals can change over time. Before using a chemical or following any chemical recommendations, the user should ALWAYS check the uses prescribed on the label of the product to be used. If the product has not been produced recently, users should contact the place of purchase or their local reseller to check that the product and its uses are still registered. Users should note that the currently registered label should ALWAYS be used.

1

Introduction

This chapter presents an outline of the book with a series of definitions to delineate clearly exactly what is meant by the term 'feedpad' and other farm facilities associated with the feedpad operation.

The main points in this chapter:

- This book presents a detailed overview on dairy feedpads, their planning and construction to their day-to-day management, from the perspective of the farmer. It is not a technical manual on how to design and build a feedpad.
- The basic components of any feedpad system can be likened to different rooms in a house, each one purpose-built and requiring specific planning, designing and over time, maintenance.
- Research and development into dairy feedpads has been conducted in Australia for over 25 years, but farmer interest has been increasing only over the last decade. In 2008 nearly 300 farmers in northern Victoria sought advice from the Victorian Department of Primary Industries' (DPIV) Dairy Farm Services.
- The DPIV has conducted dairy research and extension programs on feedpad development and management since the early 1980s.
- Before discussing feedpad designs in detail, this chapter defines exactly what the terms 'feedpad' and 'feedlot' mean and what are the differences between loafing, wintering, stand off and calving pads.

This book presents a detailed overview of dairy feedpads, from their planning to their construction to their day-to-day management, primarily from the perspective of the farmer. In recent years, much has been written on dairy feedpads, but this has been mainly from the viewpoints of effluent specialists, civil engineers, academics and building contractors. This is not unexpected given the dramatic increases in environmental concerns and accountabilities of livestock farmers regarding how they use their farm resources to make a living.

When farmers are contemplating building a feedpad from scratch (on what is called a 'greenfield' site), or upgrading an existing cow-feeding facility, naturally one expects that they would seek professional advice from a wide range of service providers. These would include:

- builders of such stock handling and management systems;
- local and state government agencies regarding existing legislation or if none is available, relevant guidelines for their construction and operation;
- suppliers of associated machinery (such as effluent pumps or mixer wagons);
- suppliers of supplementary feeds to overcome future anticipated pasture limitations;
- dairy advisers and consultants (and maybe veterinarians) to help develop the recommended management plans, feed budgets and, if necessary, environmental risk assessments;
- design engineers and concrete specialists to provide detailed drawings and building plans;
- financial advisers to assist with planning and implementing a business plan for financing its construction and day-to-day running costs; and
- dairy farmers who have developed successfully operating feedpads for their grazing cows and other dairy stock.

These service providers are likely to be specialists in their particular fields, able to provide the types of specific information required at each stage of feedpad development and operation.

Why write this manual?

Unfortunately not all farmers follow this logic of seeking out information relevant to their needs. They may simply depend on opinions from their fellow farmers who may, or may not, have had experience with these kinds of feeding and management systems. As well as providing a detailed overview of the topic, this book will complement feedback from other farmers about their experiences of intensifying their pasture-based dairy production systems through incorporating a centralised supplementary feeding station on their farm. Associated with the actual feeding is the provision of water, space, protection from climatic extremes and the removal of effluent and other feedpad wastes. This book also provides direction for farmers seeking specialist details for specific issues of feedpad construction and management.

Feedpads provide an easily visual example of intensification of farming practices that are often considered controversial by community lobby groups such

Figure 1.1   Feedpads can be simple or complex, depending on their purpose.

as those representing animal rights and environmental issues. To be comprehensive, this book then covers animal welfare as it affects cow well-being, behaviour and health, as well as effluent management and reuse.

The book does not aim to be a technical manual on how to design and build a feedpad. It does, however, contain relevant summaries from two of the key published guidelines for feedpad construction, namely those developed by the Goulburn Broken Catchment authorities in northern Victoria (SIRIC 2002) and the recently published *Guidelines for Victorian Dairy Feedpads and Freestalls* (O'Keefe *et al.* 2010). While some of this information is already over six years old, much of the civil engineering aspects has recently been updated (Dairy Australia 2008b). Guidelines tend to concentrate on statutory planning, siting and engineering design, animal health and welfare and farm safety issues with less emphasis on the day-to-day feeding and herd management of modern dairy production technology.

There are sometimes subtle differences between feedpads and feedlots. Although there is significant research on feedlots (generally beef feedlots), research in Australia on dairy feedpads is lacking. Many farmers are worried that if their feedpad becomes classified as a feedlot, they are likely to be inundated with additional legislation reducing their flexibility in system design and management. Hopefully this book will allay such fears.

This manual is written specifically for dairy farmers throughout Australia, although the key principles and practices are relevant in any dairy industry, be it another temperate developed industry, such as in New Zealand or western Europe, or a tropical developing one such as in South-East Asia. In addition, the topics covered will be equally useful for dairy advisers, industry policy makers and any group involved in training farmers (at technical farm apprentice level) or service providers (at university undergraduate level).

Although a major emphasis of this book is on herd and feeding management, rarely covered in previous guidelines and other printed material on feedpads, there are chapters with specific information on physical planning and construction of feedpads and effluent management. These chapters could be considered, in part, to duplicate those found in other guidelines such as SIRIC (2002), Dairy Australia (2008b) and O'Keefe *et al.* (2010), hence may be out of place in a book such as this. While it is hoped that virtually every dairy farmer contemplating the construction of a new feedpad or upgrading an existing one would find this book useful, it is unlikely that many would have previously perused all the technical guidelines cited above. For those who have studied the guidelines closely, we apologise for any unnecessary duplication, and trust that you can bypass these sections in the chapters while seeking new material relevant to your search. For those who are unfamiliar with the guidelines, these chapters should be of as much relevance as those on herd and feeding management. The incorporation of such chapters into this book has increased its size, but for the sake of presenting the whole story of feedpads for grazing dairy cows, please accept our justification for their inclusion.

Understanding the feedpad system

Feedpad systems are a lot more than just the actual area where the cows stand to eat. They include all structures associated with maintaining stock away from paddocks, such as the feeding area, feed ingredients storage and preparation areas for mixing rations, laneways, loafing areas and of course, effluent system. These many components to the feedpad system can be likened to different rooms in a house. The area where the cows eat can be likened to a dining room, their troughs or feeding area to a dining table. The lounging area where cows can relax, ruminate or lie down is akin to a lounge room, or a bedroom. Feed is stored in a pantry, or the feed storage area, and the mixing area is analogous to a kitchen. The hall is like the laneways through which cows walk to and from the milking shed to the paddocks to the feed area. Feeding machinery is stored and maintained in a workshop. A roof provides some degree of climate control. Alas, there is no bathroom or toilet on a feedpad because unlike pigs, cows do not prefer to move to one area to urinate or defecate. The storage and disposal (or reuse) of treated sewerage acts as a septic system.

Each 'room' requires specific planning and designing and over time, maintenance. In fact, one might even consider that building a feedpad system on a greenfield site should justify a similar degree of planning, designing and construction skills as building a house on a vacant block. Unfortunately, many feedpad systems grow over time, with little long-term planning. Few professional builders would be happy if this was the case when constructing a house, and neither should professional dairy farmers.

Long-term planning not only includes increases in the size of the milking herd and the associated facilities, it also includes considerations for managing heifer replacements and even the possibility of keeping bull calves to grow out for dairy beef. Currently all young stock may be entirely grazed, or agisted following weaning. In future years, however, heifer replacements may spend time on the feedpad as grazing pressures increase on the farm or their rearing programs require higher target live weights, to provide potentially more productive milkers and to achieve higher annual farm milk yields.

The proportion of grazed pasture in the annual diet of milkers may also vary in future years. As the cost of pasture inputs (irrigation water, fertiliser) increase or their availability may even decrease, such as costs for irrigation water due to climate changes, so too may the desired pasture mix on the farm. For example, in recent years, dairy farmers in irrigated northern Victoria have been subjected to reduced water allocations and this is leading to grazed pasture contributing less and less to milk production on these farms. The number of general enquiries for information on design, construction and management of feedpads has dramatically increased. This was first apparent following the 2002 drought and has now reached such numbers that many service providers are seeking information for their own clients. With the increased likelihood of extreme weather events, such as failed spring and autumn rains and summer heatwaves, up-to-date information on feedpad technology will become even more relevant.

In any feedpad planning, one should be aware that temporary feedpads can become permanent; a lack of investment in the early stages of planning can become costly; and poorly designed and laid out facilities may fail quickly and attract the attention of regulatory agencies.

Look around other farms for systems that may suit your own. Travel to other dairying regions and states and see what is done there. When you are visiting other operating feedpads, ask the farmers what they would do differently with the benefit of hindsight.

An outline of the manual

When discussing the diversity of variations of types of feedpads, it is important to understand exactly what constitutes a feedpad, hence Chapter 1 contains a series of definitions. Chapter 2 provides an overview of Australia's dairy industry with particular reference to how farmers need to improve their feed production technology to reduce production costs so they can cope better with the major challenges in their operating environment, such as the volatility of global world prices and climate change. We can learn much from the past, so Chapter 3 briefly provides a historical perspective on how systems of classification for feedpads have evolved since the late 1990s. This provides a framework to develop an all-encompassing classification system for feedpads in Chapter 3.

Chapter 4 discusses the physical aspects of designing and constructing dairy feedpads. As well as size and siting, the slope and surface construction material are important issues to consider. Loafing pads, free stalls and stand off pads are also discussed. Chapter 5 concentrates on nutrient management; that is, the collection and dissemination around the farm of feedpad effluent and other waste material.

Chapters 6, 7 and 8 address aspects of feeding management of dairy stock on the feedpad. To understand the basics of formulating rations for milking cows more fully, Chapter 6 introduces the principles of the nutrients contained in feeds and those required for cow performance at various stages of the lactation cycle. This is followed by a discussion on why there is so much variability in milk responses to improved feeding, which all too frequently is not comprehended fully. Partial mixed rations are discussed in Chapter 7, both their ingredients and their formulation. Agro-industrial by-products are becoming key components of many mixed rations but their incorporation requires a different approach to rations based on cereal grains. The physical requirements for feed storage are reviewed in Chapter 8 which then introduces the logistics of long-term feed planning and subsequent purchasing of the key components of all feedpad rations. The feeding of high levels of cereal grains or some by-products can lead to imbalanced diets and subsequent feeding problems. Many of these problems can be overcome by early diagnosis and incorporating rumen buffers and other feed additives into the diet. The principles of integrating feedpad technology with grazing management complete the section of feeding management.

Cow management is discussed in Chapter 9. Animal welfare is reviewed as are the two key health issues, mastitis and lameness. This chapter also addresses cow comfort, in terms of alleviating heat and cold stresses, both in terms of the required physical facilities and modified herd management. The Temperature

Humidity Index introduced in this section and also presented as an Appendix is commonly used to quantify cow comfort throughout Australia.

Chapter 10 addresses the likely effects of feedpad technology on overall farm management. The emphasis is on preparing the farm team for changes in their work practices and a review of the diversity of machinery available for mixing and feeding out the partial mixed rations.

Some of the major benefits of feedpad technology are the reduction of feed wastage and savings on fertiliser applications. As these have a direct effect on production costs, they are covered in Chapter 11 on economic aspects of installing and managing feedpads. Partial budgets are also discussed, as these provide a decision-making framework to compare the costs and benefits of alternative farm practices such as feedpads.

Chapter 12 reviews the current legislation and guidelines to be considered when contemplating a feedpad. Finally, Chapter 13 looks forward at the potential problems in feedpad technology and how they might be addressed in the future.

Several of the chapters conclude with checklists to assist farmers with their planning and preparations. All referenced sources are listed at the end of the manual and a Glossary presents the many technical terms and abbreviations used in the text. Finally, for ease of finding specific information, the Index lists all the key topics covered in the manual and their relevant page numbers.

A historical perspective of feedpad development in Australia

Although the initial feedpad research, development and extension (R, D & E) programs in Australia are now nearly 25 years old, feedpads have become fashionable in Australia's key dairy regions only over the last five years or so. Much of the early work was undertaken at the Department of Primary Industries Victoria's (DPIV) Kyabram Dairy Centre before its closure in 2009. To provide some indication, dairy effluent extension specialists at DPIV have been advising northern Victorian dairy farmers on feedpad development since the early 2000s with the number of farmer enquiries rising dramatically in recent years, as shown in Figure 1.2.

Rainfall inconsistencies, high summer temperatures and increasing environmental regulations are the key reasons for this reinvigoration of feedpad interest.

Much of this initial interest in feedpads arose from changes in dairy production systems. Australia's first commercial dairy feedlots were established in the late 1970s and this led to DPIV Kyabram instigating a research program into 'housed cow production systems' in the early 1980s. Commercial dairy production in southern Australia at that time was based on grazed pastures supplemented with conserved fodder (pasture hay and silage), while some concentrates were fed to

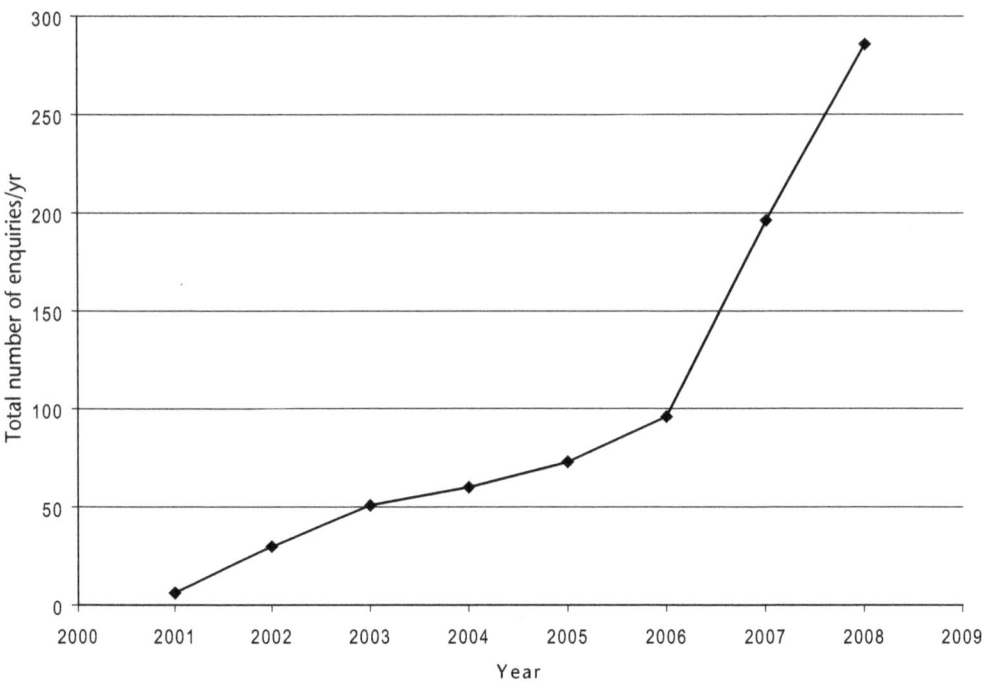

Figure 1.2 Number of feedpad enquiries each year from dairy farmers to DPIV dairy effluent specialists in northern Victoria.

grazing dairy cows in northern Australia. Concentrate feeding slowly increased since then while as stocking rates increased, more conserved forages were fed.

Dairy farmers in southern Australia supplied the manufacturing milk market primarily, and had to develop low-cost pasture-based production systems based on spring calving cows. NSW and Queensland farmers, however, supplied a more lucrative liquid milk market. They had to produce milk, hence calve, year round, thus required more concentrates to supplement their lower quality pasture base. This all changed in 2000 when Federal Government deregulation effectively removed state boundaries and Australia's dairy farmers were subjected to a single series of price signals. There were still differences in farm management between dairy farmers predominantly supplying domestic or export markets. The 2002 and 2006 extreme drought year also led to changes in the feedbase on the nation's dairy farms, as farmers imported more supplements to meet the nutrient shortfalls of reduced pasture growth.

Typical levels of supplementary feed (in tonnes of dry matter per cow per year) in Australian dairy herds (Ashton and Mackinnon 2008) were:

- 1980–1990: 0 to 0.7 t concentrate and 0.5 to 1 t conserved forage
- 1990–2000: 0.5 to 1.8 t concentrate and 0.5 to 1 t conserved forage
- 2000–2009: 1.5 to 2.5 t concentrate and 1 to 2 t conserved forage

Table 1.1 Timelines for feedpad development in Australia.

Years	Research and extension	Key publications
1980–1985	Dairy feedlot research at DPIV Kyabram	
	Dairy feedpad research at DPIV Kyabram	
1985–1990	Dairy feedpad research at DPIV Kyabram	Moran (1989) feed pad sheet
	First feedpad field day at DPIV Kyabram (1988)	
	Beef feedlot/feedpad research at DPIV Kyabram	
1990–1995	Beef feedpad research at DPIV Kyabram	
	Dairy feedpad research at DPIV Kyabram	
	Dairy feedpad and feedlot research in Queensland	
1995–2000		Moran (1996) forage conservation book
		Davison and Andrews (1997) feedpad booklet
2000–2005	Feedpad field day and farm tours at DPIV Kyabram (see Figure 1.3)	SIRIC (2002) feedpad guidelines
		Kaiser et al. (2004) silage book
2005–2010	Feeding systems farm walks in northern Victoria	Dexcel (2005) NZ feedpad booklet
	Flexible feeding systems field days throughout Australia	Dairy Australia (2007a, 2007b) fact sheets
	Service provider workshops on feedpads in Victoria	McDonald et al. (2008) feedpad booklet
		Dairy Australia (2008b) effluent database
		O'Keefe et al. (2010) Victorian guidelines

The timelines of developments in feedpad technology since the early 1980s are presented in Table 1.1.

Definitions relevant to feedpads and feedlots

Before discussing feedpad designs in detail, it is worth noting exactly what the terms 'feedpad' and 'feedlot' mean and what are the differences between loafing, wintering, stand off and calving pads.

Figure 1.3 Field days are an effective way to improve awareness of the costs and benefits of feedpads.

A *feedpad* is defined as that part of a dairy farm that is used for supplementary feeding of stock on an area of land that is formed, surface or stocked at a rate that precludes vegetation.

This is generally a confined, yard or laneway area in which feed and/or water can be provided. Stock are held and mechanically fed for the purpose of milk production of milking cows and/or growth of young stock, and for protection from adverse environmental impacts such as wet, cold or hot conditions.

The term 'feedpad' usually incorporates not only the pad, but also the associated supplementary feeding system such as the feeding area, water troughs, feed storage, laneways and effluent collection, storage and/or reuse.

The pad is that area of land formed, surfaced or stocked, typically raised to assist drainage, to provide a dry surface for feeding the stock and to minimise wear from stock and machinery. It can also include a *loafing pad* or lounging area where the stock can ruminate, seek shade and if contained for lengthy periods, lie down. This is also known as a *stand off pad*. Feedpads are sometimes called *wintering pads*, which serves two purposes, namely protecting wet pastures and providing supplementary feeding.

Another term used for a feedpad, although primarily with beef cattle and sheep farmers, is a *stock containment area*. This is defined as an area on the farm set aside to assist with stock management during adverse climatic conditions, prolonged drought and in times of emergencies such as fire or flood. They are usually set up with permanent feeding and watering facilities, so are effectively feedpads. Some government agencies provide financial assistance for their establishment, which can be used to purchase fencing, gates, troughs, piping, tanks and pumps.

Farmers may decide to feed out conserved forages (hay or silage) to cows on a paddock, rather than in a confined yard. They may be fed in a paddock that has either been grazed previously or is yet to be grazed or in a sacrifice paddock. The term *sacrifice paddock* then describes a paddock that will not be used for grazing in the short term, such as one that has been removed from the grazing rotation and destined for pasture renovation or for ploughing to grow a fodder crop. Until it is ploughed, its purpose is purely to provide space for cows to be fed supplements and as a loafing area. Wherever possible, farmers should avoid using paddocks adjacent to streams or rains, to reduce the risk of soil and nutrient movement to these waterways.

A *calving pad*, on the other hand, is an area on dairy farms specifically designated for calving down the cows. It is constructed to provide a warmer, drier option to the paddocks and in close proximity to yards to enable round-the-clock access for the observation and care of cows and newborn calves. Calving pads may incorporate subsurface drainage and are typically covered with some absorbent organic bedding such as rice hulls, straw or sawdust.

Feedpads differ from *feedlots* in that cattle in feedlots are not given any access to pastures for grazing. Feedlots can then be defined as land on which cattle are restrained by pens or enclosures for the purpose of intensive feeding. There is considerable government legislation regarding beef feedlots, whereas dairy feedpads are far less controlled. In essence, a feedpad is a feedlot with the gate left open.

To allow for extremes in feedpad size, legislations or guidelines developed by some local government agencies may only consider cases within a minimum and maximum size. For example, SIRIC (2002) consider feedpads for less than 50 head or for more than 5000 head as being 'anomalous sized feedpads' which have a simpler set of guidelines (<50 head) or require approval by a different agency (>5000 head).

In the loafing area, stock can be managed in loose housing or in free stalls. With *loose housing*, stock are free to lie down anywhere on the pad, in contrast to *free stalls* where cows are allocated specific areas to lie down which they may enter and leave at will. *Free stall sheds* (sometimes called barns) are then the structures in which cattle are provided with individual free stall cubicles. Other features of these systems include roofing (at least over the cubicles), regular cleaning of the laneways and frequent collection and reuse of the manure and bedding.

A frequently used term in feedpad technology is *buffer distance*. This defines the distance between the feedpad and existing housing, land zoned for residential or urban purposes, or other sensitive uses. The criteria used to determine buffer distances include potential noise or odour, which relate to the type of feedpad, stocking density and length of time cows spend on the feedpad; this is quantified in terms of Dairy Cow Units which are discussed later in Chapter 4. Buffer distance also relate to the frequency and method of cleaning the feedpad, the type of residences nearby (single house or town), the terrain (whether the feedpad is on flat land, in a valley or on a hill) and degree of vegetation (heavy, light or negligible tree cover). As it is recommended that a dairy effluent pond should not be situated within 300 m of a neighbouring residence, this can be used as an initial guide to acceptable buffer distance for a feedpad. Further details of considerations and an example of the calculations involved are presented by SIRIC (2002).

2

An overview of Australia's dairy industry

This chapter provides a snapshot of Australia's dairy industry with particular emphasis on adoption of improved production technology.

The main points in this chapter:

- Over the last 15 years, the number of dairy farms in Australia has nearly halved, yet milk production has risen by over 40%.
- Total cow numbers and average milk yield per cow have increased, leading to a moderate productivity growth of 1.2% per year.
- Many external changes in their operating environment have led Australian dairy farmers to modify their farm management practices through adopting new technologies such as providing more controlled levels of supplementary feeds on feedpads.
- The more profitable farmers have adopted these technologies more readily.
- Between 1990 and 2007, farm gate milk returns steadily increased from less than $3 to over $4/kg milk solids (MS); however, during 2008, they reached a peak of over $5/kg MS then within 12 months fell to less than $3/kg MS.
- Over the last 18 years, feed barley prices have varied from less than $100 to more than $350/t.
- In 2009, the milk:feed price ratio was at one of its lowest values for the last 20 years, indicating a severe cost-price squeeze for Australia's dairy industry.
- In 2006, about 18% of Australia's dairy farmers had feedpads, with more on the larger farms (for example, 31% on farms running more than 500 cows). This amounted to a total of 1840 feedpads on Australian dairy farms.
- Assuming many of the current inquiries for feedpad information are from farmers installing new pads, there could be 2000 feedpads in operation by 2010.
- Following a national survey of various dairy production systems, the number of farms already using or planning to install feedpads is 1767, with the highest number (601) in the irrigated pasture-based regions of northern Victoria and the Riverina.

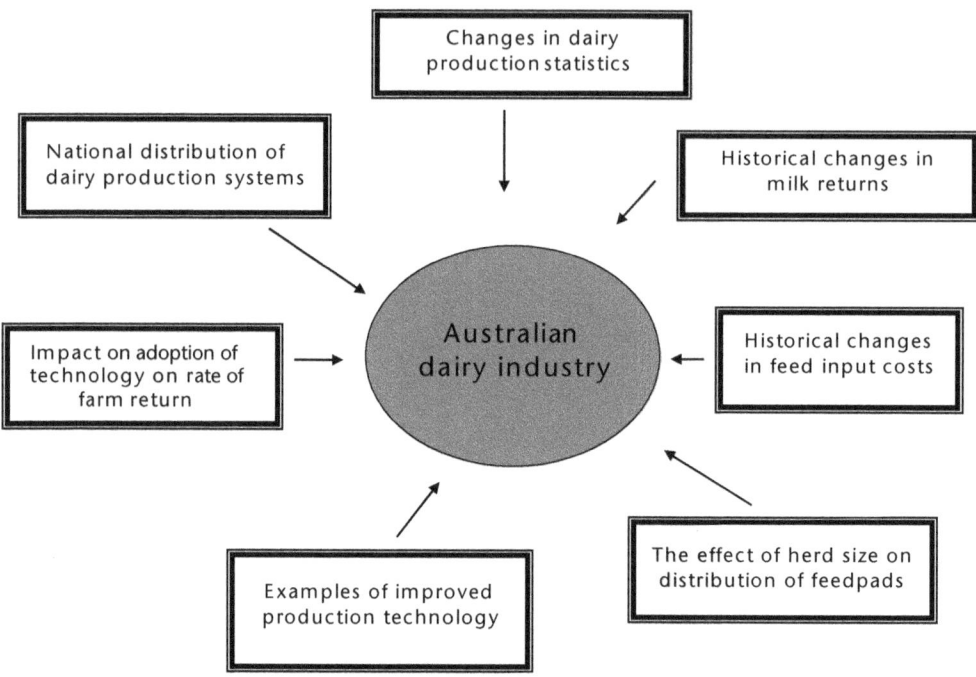

Figure 2.1 A snapshot of Australia's dairy industry.

Feedpads are just one example of advanced feeding technology used by dairy farmers around Australia. They have evolved into multi-use facilities, split calving herds, hospital areas, weather stand off, shade areas; that is, more than just a feeding out area. Virtually every dairy farmer now seeks to provide their stock with supplementary feeds in addition to grazed pasture, and being cost-conscious, they need to minimise feed wastages. During the 1970s and 1980s, dairy farmers, particularly those in southern Australia, were content to feed conserved roughages, usually hay, out in the paddock, in addition to small amounts of cereal-based concentrates in the milking shed. As feed costs rose, more physically efficient feeding systems had to be sought to reduce wastages. In addition, as grain feeding became universal, and the quantities fed increased dramatically, more nutritionally efficient feeding systems had to be sought to ensure better milk responses to these supplements.

This chapter helps place feedpads in perspective in Australia's dairy industry through a close study of a recent review of improved technology and management practices.

Australia's dairy industry is exposed continually to many external pressures such as changing climate, government policies and competition for global markets. In recent years, the industry has been adversely affected by a number of these factors which have placed downward pressure on farm incomes, such as drought,

Figure 2.2   Hay has been the traditional forage supplement of dairy cows in southern Australia.

low or zero allocations of irrigation water in some regions and high prices for fodder and feed grains (see Figure 2.1). This has increased the volatility of farm profits over the last decade, which has been exacerbated by large fluctuations in world prices for traded dairy products.

Such changes in their operating environment are major incentives for dairy farmers to modify their farm management practices by adopting a variety of new technologies to improve their farm performance and profitability. These practices have led to gains in milk production per cow as well as increased labour productivity and farm size. In addition to herd health, shed management and dairy cattle genetics, there have been changes in pasture management and supplementary feeding practices. Constructing feedpads for grazing milking cows is just one example of new production technology. In addition, farmers unable to maintain profitability have left the industry allowing the remaining farmers to increase both the size and the intensity of their operations.

Recent changes in dairy production technology

Over the last 15 years, the number of dairy farms in Australia has nearly halved (from 14 000 in 1994 to 7900 in 2008) yet milk production has risen by over 40% as total cow numbers and average milk yield per cow increased. Between 1988 and 2005, this amounted to a moderate productivity growth of 1.2% per year.

Figure 2.3   Growing forage maize for silage has been a very successful innovation for Australia's dairy industry.

The average dairy herd size (which includes young stock as well as miking cows) in 2009 was 274 head, with 9% of the farms (comprising 25% of the national herd) having herds with 500 cows or more (Dairy Australia 2009c). This was associated with a small increase in stocking rate (1.4 in 1999 to 1.6 head/ha in 2006), whereas annual milk production per cow has increased from 3980 L (in 1991) to 6171 L (in 2009). Purchased feed was the single biggest cost item on farm, in 2009 comprising 39% of dairy farm cash costs, up from 24% a decade ago. For many farmers, this represents the greatest challenge they will face in the future.

Ashton and Mackinnon (2008) reported on changes in production technology over the last 15 years (from 1991 to 2006, based on annual surveys of about 300 dairy farms) with some of their data presented in Table 2.1. In 2006, farmers fed nearly three times more concentrates than in 1991. The breakdown of concentrate feeding indicates a large increase (from 1991 to 2006) in the quantity of by-products fed. Other indicators of improved farming technology include greater soil testing (conducted by 58% of farmers in 2006) and increased attendance at farmer training courses, particularly those related to herd nutrition, feeding and pasture management.

The vast majority (97%) of farmers now feed concentrates to their grazing cows, generally in the milking shed. Since not all have bail-feeding facilities,

Table 2.1 Changes in concentrate feeding practices on Australian dairy farms (% of farms, t/farm/yr) and in farms purchasing hay or silage (%).

	1991	1999	2006
Farms feeding concentrate (%)	81	91	89
Farms with no bail feeder (%)	25	17	14
Farms purchasing hay/silage (%)	52	54	68
Self mixed concentrates (t/yr)	11	15	41
Purchased concentrates (t/yr)	38	74	149
Grain (t/yr)	53	124	147
By-products (t/yr)	6	23	41
Total (t/yr)	**109**	**236**	**377**

(Source: Ashton and Mackinnon 2008)

concentrates are sometimes fed, while conserved forages are usually fed at pasture or on a feedpad. Feeding losses can be quite large when forages are fed in the paddock, leading to economic inefficiencies. When hay or silage costs increase, as they invariably do during supply shortages such as experienced in recent droughts, these losses can increase farm input costs dramatically, markedly cutting farm profits.

In 2006, the better-performing farms achieved greater profits through higher milk yields, lower cash costs and higher cash income per cow, thus leading to higher rates of return to farm investments. Seasonal milk producers, whose cows are generally mated to calve and commence lactation in the period of peak pasture availability, were able to generate above-average farm cash income per cow, compared to year-round producers.

Even though cash costs per cow and farm cash income per cow do not differ greatly with size of the milking herd, the larger farms tended to generate higher rates of return on total farm capital by spreading overhead costs, such as purchasing feedpad technology, over more units of output.

Examples of improved production technology

New investments are an important means of boosting farm productivity and incomes, with productivity growth providing better prospects for business viability in the longer term. In addition to acquiring land, higher farm incomes have allowed farmers to expand the scale of their operations.

The shift towards larger farms and increased intensity, such as higher stocking rates and more intense feeding practices, then contributed to increased production over the last two decades. In particular, pasture improvement, the expanded use of fodder harvesting technologies and increased use of grain, concentrates and fodder have contributed to these production gains. In 2006, over 90% of farmers fed concentrates, with average feeding rates increasing from 0.9 t/cow in 1991 to 1.8 t/cow on 2006. Farmers feed more concentrates to achieve higher milk yield per cow, to overcome shortfalls in pasture quality and quantity and to assist in grazing management.

Another indicator of intensifying production systems is the increased purchase of hay or silage. This rose from 52% in 1991 to 68% of all dairy farmers, in the drought year of 2006. This is an additional forage source to that conserved on the farm, a normal practice for 70–80% of farms, which has not changed greatly over the last 15 years.

In addition to formal herd nutrition and feeding management courses, attended by 30% of farmers in 2006, farmers seek a wide variety of sources of information to manage their dairy enterprises better. Farmer discussion groups provide the opportunity to see improved technology in practice. In 2006, about half of all dairy farmers indicated management advice helped them increase the profitability of their farm businesses through improved supplementary feeding, herd nutrition and better pasture and forage management practices. One surprising finding reported by Ashton and Mackinnon (2008) was that in 2006, around two-thirds of dairy farmers used computers for farm or herd management decision making. This is surprising, given the average age of dairy farmers and the interest in computer technology displayed by similar aged people in the general community.

Another finding relevant to intensification of dairy systems is the dramatic reduction in farmers discharging dairy effluent directly onto paddocks, which has fallen from 52% in 1991 to only 20% in 2006. This has been attributed to responses to environmental concerns and changes in state-based regulations leading to more farmers using ponding systems.

Recent changes in dairy farm costs and returns

Farm gate milk returns for dairy farmers in southern Australia are largely influenced by fluctuations in the export market for processed dairy products such as whole milk and skim milk powder and butter and cheese. Figure 2.4 presents average milk returns, not adjusted for inflation, expressed in $/kg milk solids (MS) for Murray Goulburn suppliers over the last two decades. On a milk volume basis, $4.50/kg MS is equivalent to 33 cents/L while $7.00/kg MS is equivalent to 50 cents/L.

The sudden deterioration in the health of the global economy during late 2008 had a dramatic effect on export milk prices. This led to farm gate milk returns over the last five months of the 2008 season dropping by 30–40% from previously

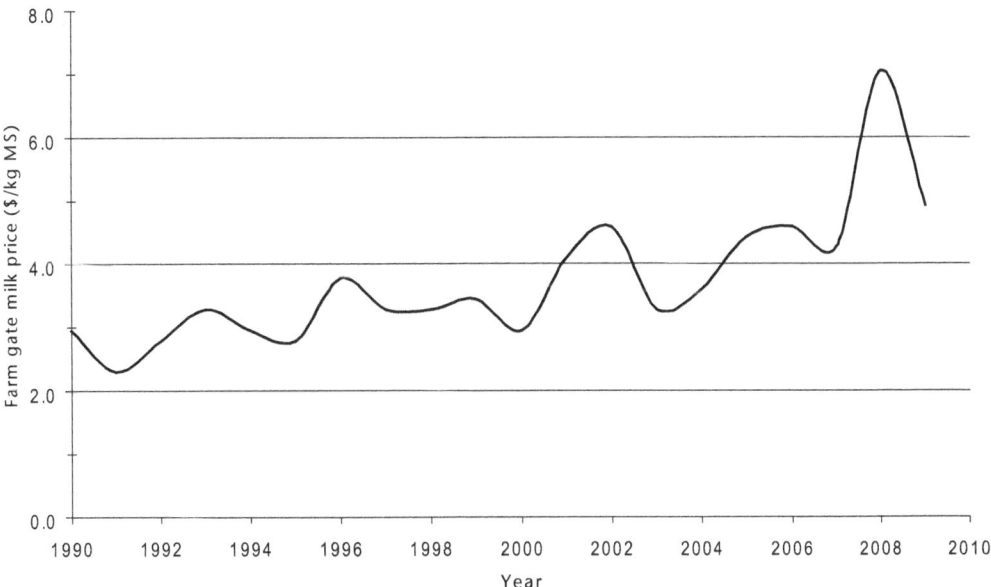

Figure 2.4 Average farm gate milk returns for Murray Goulburn suppliers in southern Australia from 1990 to 2009 (Dairy Australia 2009b).

announced opening prices (see Figure 2.4). Opening prices in the 2009 season were $3.50–$4.00/kg MS, equivalent to 26–29 cents/L.

Since feed contributes over 50–60% of total costs of milk production, the costs of feed inputs has a major influence of farm costs. The price of feed barley can be used as a key indicator of concentrate costs, as presented in Figure 2.5.

With increasing quantities of forages purchased by dairy farmers, the price of lucerne hay is a valuable measure of forage costs. Figure 2.6 presents data over the last five years for lucerne hay purchased in northern Victoria.

Industry analysts use the milk-to-feed price ratio as an indicator of the relative cost of milk production. Figure 2.7 presents historical data of the milk:feed (barley grain) cost ratio over the last two decades. A value of 1.0 indicates that dairy farmers would receive the monetary value of 1 kg of barley grain per litre of milk produced. Milk responses to cereal grain supplements will be discussed in Chapter 6.

These four graphs indicate clearly the cost price 'squeeze' being experienced by Australia's dairy farmers during 2009. Although not as severe as 2003 (from Figure 2.7), dairy farmers are seeking ways to improve their milk production economic efficiency, particularly as far as feed conversion is concerned.

One indication of the reduced availability of pastures for dairy farmers, hence their significant exposure to high grain and fodder prices, was that during 2008, over 75% of the dairy farmers in northern Victoria and the Riverina maintained

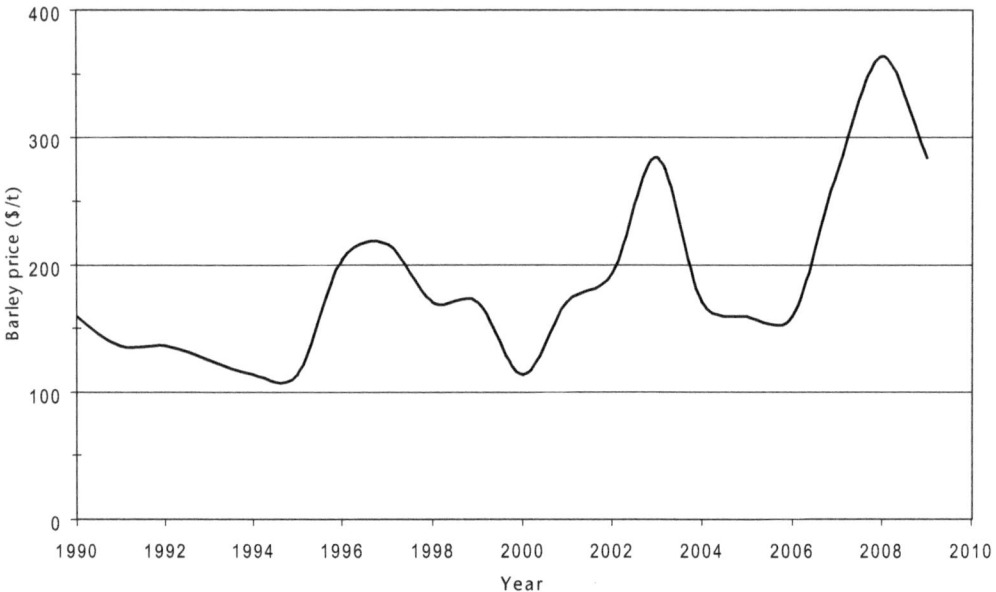

Figure 2.5 Long-term trends in the price of barley grain (Dairy Australia 2009b).

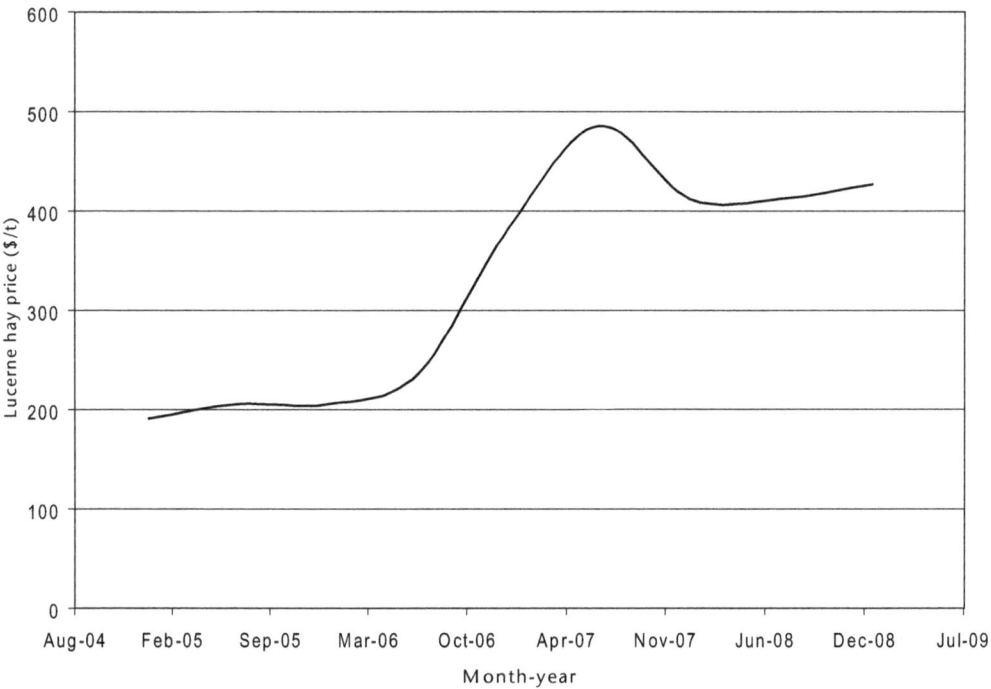

Figure 2.6 Changes in the price of lucerne hay purchased in northern Victoria over the last five years (Dairy Australia 2009b).

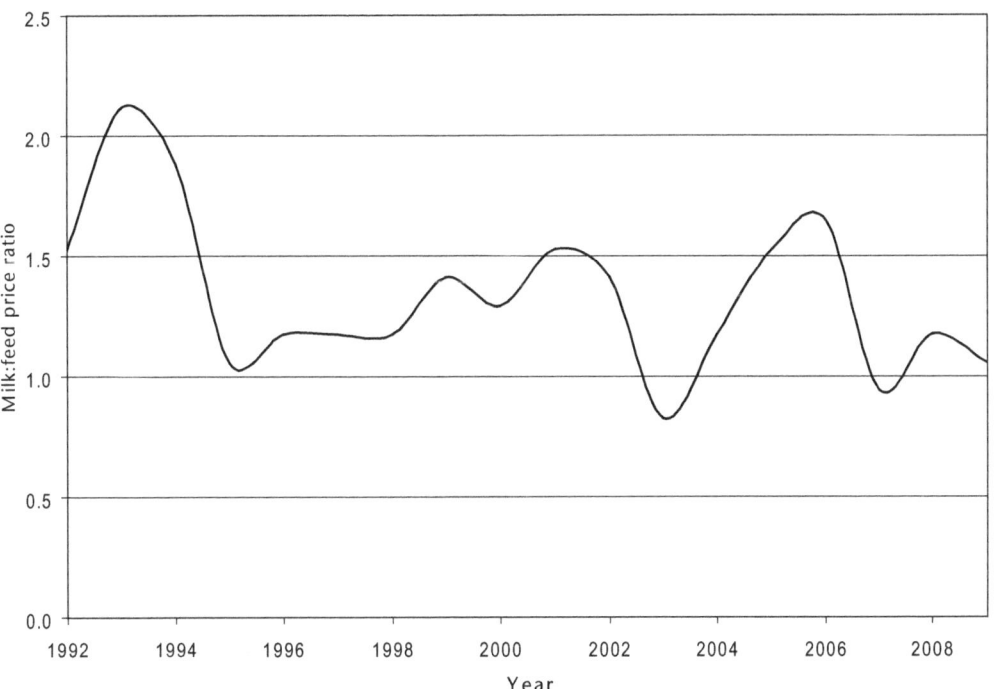

Figure 2.7   Long-term trends in the milk:feed (barley grain) price ratio (Dairy Australia 2009b).

their milking herds exclusively on purchased feeds for an average of 6.3 months of that year (Dairy Australia 2009b). In a region where dairy farming has traditionally been based on irrigated pastures, this highlights the dramatic impact of reduced water allocations on the cost of feeding stock during the 2008 drought year. Many of these farmers have remained committed to, or have changed their operations to, higher cost production systems with an expectation of the sustained high milk prices of early to mid-2008. The large importations of expensive bought-in feed in association with the sharp fall in milk revenue, however, has left them exposed to higher feed and overhead cash costs hence markedly reduced farm profit margins. Clearly, as many are planning to utilise the latest in feedpad technology, this manual has considerable relevance to this dairying region.

Performance by rate of return

Ashton and Mackinnon (2008) used 2006 survey data of 291 farms throughout Australia to explore relationships between farm profitability and use of technology. Farms were ranked according to their individual rates of return to capital and management, then grouped into quartiles (that is, groups of 25%), with comparisons made in Table 2.2 between those in the bottom (0–25%) and those in the top (75–100%) quartiles.

With regards to various aspects of feeding technology, more of the most profitable farmers intended to attend training courses or to improve current farm technology. Of the most profitable farmers surveyed, nearly double the proportion of those least profitable, currently participate in farmer discussion groups. The majority of least profitable farmers either have no or only a manual bail-feeding system in their milking shed in contrast to the most profitable farmers, who mainly have mechanised or computerised bail-feeding systems. Nearly half the least profitable farmer group allow their shed effluent to run off directly into the paddock.

Higher farm profits are then associated with improved supplementary feeding practices and a greater awareness of the need to know how to reduce feed nutrient losses through better feeding management. The more profitable farmers are those who can provide their milking cows with sufficient amounts of high quality feeds, both basal forages such as grazed pastures or conserved hay and silages and supplements, such as formulated concentrates and/or agro-industrial by-product feeds. Minimising feed wastage and balancing the nutrient requirements of grazing cows are the two main avenues available to Australian dairy farmers to optimise feed nutrient utilisation which leads to improved feeding efficiencies and lower milk production costs.

Survey of effluent management on feedpads

Anonymous (2006) reported on a survey of 1200 dairy farmers, representing 12% of Australia's dairy farmers at the time, on their current natural resource management practices and compared it with data from a similar survey conducted in 2000. The sampling procedure was based on the number of dairy farmers in each major dairying region. The survey included questions on effluent management of feedpads. In 2006, the average herd size was 233 cows grazing on an average of 152 ha.

The proportion of farmers with feedpads had not greatly changed over six years, from 19% in 2000 and 18% in 2006. From Table 2.3, this proportion increased with herd size from 15% of farms with less than 150 cows to 31% of farms with more than 500 cows. Since 2000, the proportion of respondents with feedpads who managed feedpad effluent by collecting it in a pond increased from 24% to 43% as did those using a sump and dispersal system. Consequently, a lot less farmers drained their feedpad effluent directly onto paddocks, from 56% in 2000 to 24% in 2006.

Extrapolating these data to the entire industry, which in 2006 totalled 10 000 dairy herds, the number of Australian dairy farms with feedpads amounted to

Table 2.2   Utilisation of technology on 291 Australian dairy farms in relation to their rates of return to capital and management in 2006: comparing farms in the bottom quartile with those in the top quartile.

% of farmers intending to or already using improved production technology	Bottom 25%	Top 25%
Intended training course within 12 months:		
Herd nutrition	10	38
Concentrate and feed management	18	29
Fodder conservation	13	20
Currently participates in farmer discussion group	33	58
No training likely	48	29
Intended management changes:		
Install new feeding system (including feedpads)	2	17
Improve supplementary feeding and herd nutrition	22	39
Increase grain/concentrate feeding	4	17
Increase use of farm consultant	1	7
Farms that currently soil test	27	86
No significant change likely	40	23
Bail feeding type in milking shed:		
None	21	8
Manual	47	15
Mechanised	31	59
Computerised	1	17
Effluent disposal system from dairy:		
Runoff into paddocks	49	4
Pump and spray	4	19
1 pond system	25	28
2 pond system	10	40
Other systems	11	8

(Source: Ashton and Mackinnon 2008)

about 1840. Assuming many of the enquiries for feedpad information (see Figure 1.3 in the previous chapter) are from farmers planning new feedpads, the number of dairy feedpads in operation could well exceed 2000 by 2010.

Table 2.3   Distribution of feedpads in Australia in 2006 based on herd size, based on a survey of 1200 farms.

Herd size (cows)	% farms	% with feedpads	Number of feedpads in surveyed farms	Number of feedpads in Australia
<150	36	15	65	541
150–300	43	18	93	775
301–500	15	23	41	342
>500	6	31	22	183
Total	–	18	221	1841

(Source: Anonymous 2006)

Survey of dairy feeding systems throughout Australia

In 2009, the average grain or concentrate-feeding rate was 1.5 t/cow/yr. Only 6% of the farmers fed less than 0.5 t/cow/yr, 51% fed between 0.5 and 1.5 t/cow/yr, 23% fed from 1.5 to 2.0 t/cow/yr while 13% fed more than 2.0 t/cow/yr (Dairy Australia 2009c). Larger herds fed at higher rates than did smaller herds. The unfavourable weather conditions led to increasing periods during the year when cows were not grazed at pasture, depending entirely on hand-fed forages and grain/concentrates. This averaged 4.8 months/yr throughout Australia, but as high as 6.3 months/yr in northern Victoria.

Little (2009a) reported on a survey of 1000 dairy farmers throughout Australia, in which their feeding systems were classified into one of five distinct categories. These were based on the level of concentrates fed, whether it was fed entirely in the milking shed or some was also fed on a feedpad, and the number of months per year the herd was fed entirely on supplementary feeds, i.e. there was no grazed pasture. Concentrates included both 100% cereal grain and other formulations incorporating grain. The five systems were:

1. Low grain: where cows were offered grazed pastures plus other forages in the paddock together with up to 1 t concentrates/cow/yr fed in the milking parlour.
2. Moderate to high grain: where cows were offered grazed pastures plus other forages in the paddock together with more than 1 t concentrates/cow/yr fed in the milking shed.
3. Partial mixed ration (PMR): where cows grazed pastures for most of the year and were fed a PMR with or without additional concentrate feeding in the milking shed.
4. Hybrid system; where cows grazed pastures for less than nine months each year and were fed forages and concentrates incorporated into a PMR.
5. Total mixed ration (TMR); where cows were zero grazed hence continually housed and fed their forages and concentrates entirely as a TMR.

Each major dairy region or state was surveyed and the findings are summarised in Table 2.4. A sixth category of 'Other' allows for production systems that cannot be classified as one of the five systems described above.

Assuming Systems 3, 4 and 5 are those where feedpad technology is already adopted, the sum of these three systems ranges from a high of 36% in NSW to a low of 12% in western Victoria, with an average of 20% for the Australia's entire population of dairy farms. Based on the number of dairy farms in each region (Dairy Australia 2009a), this equates to 1767 dairy farms where feedpad technology is relevant. These numbers do not even take into account any of the 'Other' feeding systems (which number 289).

Table 2.4 Distribution of dairy feeding systems in Australia. See text for definitions of feeding systems.

Region	System 1 (%)	System 2 (%)	System 3 (%)	System 4 (%)	System 5 (%)	Other (%)	Systems 3, 4 & 5 (%)	Number of System 3, 4 & 5 farms
Northern Victoria/Riverina	15	49	12	13	6	5	31	601
Western Victoria	34	51	9	2	1	3	12	184
Gippsland	31	56	13	1	0	1	14	241
New South Wales	17	46	20	7	9	1	36	353
Queensland	17	50	23	2	8	0	33	221
South Australia	10	69	7	10	5	0	22	70
Western Australia	17	65	12	5	0	2	17	30
Tasmania	56	29	15	0	0	0	15	67
Australia	**25**	**52**	**13**	**5**	**2**	**2**	**20**	**1767**

(Sources: Little 2009a and Dairy Australia 2009a)

Farmers choose to invest in Systems 3, 4 and 5 for a variety of reasons. Some of these, listed by Dairy Australia (2009c), were to achieve higher feed intakes and better control over diets; to utilise cost-effective by-products; to reduce levels of feed wastage; to provide passive or active cooling to cows in hot weather to sustain daily feed intakes and milk production; and to control wet weather damage to pastures.

In recent years, there has been a gradual shift in many regions along the spectrum from System 1 towards 5, as the dairy industry adapts to changing climatic and farm business environments. Little (2008) stressed that each dairy system can be profitable under Australian dairying conditions, given the right mix of management, milk returns and input costs. It is not simply a case of one system being better than the others. Currently there is a wide range of productivity and profitability being achieved with each of these systems. This clearly indicates there is room for improvement on many farms. This includes the development of more efficient and cost-effective supplementary feeding systems.

The key issue is then for each individual farmer to decide which system suits them best then to make the most of the system selected. History shows that feeding systems evolve as the farm business evolves. Because a system that is ideally suited for 2009 may need to be modified in future years, it is important for farmers to be aware of the other options available. Moving along the spectrum of systems requires increasing capital outlay in feeding infrastructure and equipment, to provide greater control and reduced feed wastage. The economic implications of this are discussed in Chapter 11.

3

The role of feedpads in pasture-based dairy farming

This chapter provides a classification of types of feedpads on dairy farms.
The main points in this chapter:

- Feedpads can provide a wide range of benefits such as reducing feed wastage, reducing climatic stress, improving cow performance and better pasture utilisation. All this can lead to higher farm profitability.
- Feedpads form part of the risk management strategy on dairy farms in that they provide the flexibility to provide an efficient system to feed supplements in addition to grazed pasture.
- Types of feedpads can be categorised using a variety of criteria, with an all-encompassing classification scheme using four criteria: feedpad surface, feeding system, lounging area and type of overhead over.
- The system of effluent management, the machinery for feed preparation and delivery and the type of feed storage could provide three additional classification criteria.
- With many feedpads evolving over time, there are a series of trigger points for instigating various farm plans. These can help prioritise the work, be used to formulate budgets, stage development and then track and improve farm performance while planning for unexpected events.

This book describes the planning, design and management of feedpads on a grazing dairy farm. Prior to wanting this information, farmers must firstly be aware of the benefits of these innovations, and secondly decide on their potential economic efficiencies for their particular production system, be it tropical or temperate, dryland or irrigated (see Figure 3.1).

Feedpads can be utilised on a regular basis, or just when required. They are frequently used during wetter colder months, when access to pastures and pasture

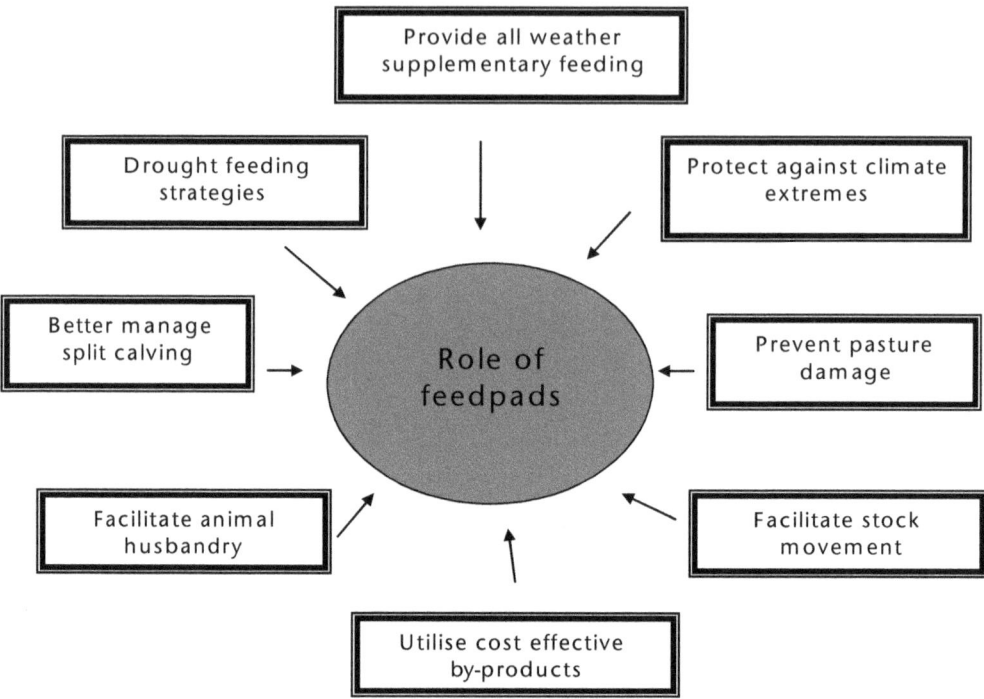

Figure 3.1 The role of feedpads on pasture-based dairy systems.

availabilities are limited, or during summer to reduce heat stress. Farmers usually provide daily access to feedpads either just prior to milking when cows are collected from the paddocks, or following milking prior to offering the cows pasture for grazing. In more recent times, due to continuing drought and more extreme climatic conditions, the use of feedpads for extended periods is becoming more common as farmers adjust their management practices to ensure herd comfort and performance are paramount. The opportunity to utilise shaded areas or more elaborate cooling systems, such as fans and sprinklers, has enabled farmers more management flexibility to sustain production during the summer months.

Planning feedpad development

Their primary aim is then to provide all-weather supplementary feed to improve dairy herd performance (milk, fertility, live weight gain). This usually occurs during periods of shortages of pasture or when pasture quality is sub-optimum. Additional uses include protection from extreme climate stresses (hot or cold); prevention of pasture damage during renovation, over winter or following lengthy wet periods at any time of the year; preventing damage to laneways sometimes used for supplementary feeding; facilitating stock mustering as cows voluntarily walk to feedpads to seek supplementary feed; and facilitating animal husbandry practices, such as with calving cows. Managing a split calving herd, such as better feeding of autumn calvers over winter becomes easier, and farmers will notice a shift towards contract services to reduce equipment and/or labour when using a feedpad. They can be used as part of a farm's drought strategy to minimise the adverse effects of seasons with poor water allocations in irrigation areas or poor pasture growth in dryland regions, and used as a quarantine facility for sick or injured stock.

Why develop a feedpad?

There are many reasons to want a feedpad, such as to establish a centralised area for supplementary feeding before or after milking. They can help reduce unnecessary wastage and spoilage of fodder and supplements; reduce the fouling and trampling of pastures where hay is normally fed out; reduce the spread of weeds, such as barley grass, brought onto the farm in purchased hay; reduce the likelihood of pastures pugging, hence needing renovations, during the wetter months of the year; and reduce animal stress by providing shade and shelter in extreme climatic conditions such as heat, wind and heavy rain events. A feedpad can assist in the management of fluctuating water allocations, which directly affect pasture production. It can help reduce feet problems and improve cow travel. Feedpads provide a facility to enable specialised supplementary feed rations to be incorporated into the cow's diet. They can help collect effluent for more even applications to grazed pastures. They facilitate an increase in the range of supplements than can be fed out. They provide a better environment for sick and injured cattle. Workloads on farms with appropriate machinery can be reduced, and they can provide a short-term contingency measure during drought. They help minimise losses in conception rates, milk protein and fat contents, live weight losses, somatic cell counts and clinical mastitis. Many farmers just want one, due to peer pressure, unfortunately still an all-too common incentive.

Introducing a feedpad onto the farm often leads to other changes in the production system. These can include increasing farm milk production because of improving utilisation of existing feed supplies. Is the existing milk vat of sufficient capacity? A feedpad may increase the need to learn new management skills, such as ration formulation. They can change the amount of purchased feeds; cause farmers to spend more money upgrading other farm infrastructure, such as a larger effluent system, bigger tractors or feed out carts. They can also increase the risks to the farm business through larger loans and reduced farm equity; increase labour and the skills of existing staff.

Considerations when planning feedpads

To be effective, a feedpad needs to be considered as integral to a farm and its management strategies by providing a multi-purpose facility to complement existing year-round management. Farmers should plan for a feedpad by asking why is there a need for a feedpad and what will it be used for? What are the perceived benefits of a feedpad? What is the most suitable type of feedpad and how big should it be? How much money should be invested and over what time period will it be constructed? What is the long-term strategy of the farmer, such as years left to dairy farm, succession plans and target herd size? What is the cost of the feedpad system compared to the long-term lost pasture production if persisting with existing pasture management? What are the likely fluctuations in stock feed prices and milk returns? What investments are required in addition to the physical construction; for example, machinery needs? What additional management skills will be required to make best use of a feedpad? Increased handling and concentration of stock on a feedpad often leads to animal health issues resulting in more lameness, declining milk quality and elevated risk of stock disease transfer. It is worth remembering that the greater the investment, the more multi-purpose the facility needs to be.

Feedpads often evolve as part of an intensification program. An impulsive approach to deciding on their design and level of investment can be dangerous. As they are not cheap, such decisions should include long-term planning on changes in farm layout, herd size, production system (such as incorporating purchased forages or agro-industrial by-products), target herd milk production and changes in approaches to herd management. It is essential to seek current advice on local government and municipal planning regulations or codes of practice, as these tend to differ between regions. Even if there are no actual regulations in place, planning should be based on guidelines which have been developed along the following fundamental principles:

- No excess amounts of nutrients, salts, chemical, debris, noise, pests, insects, microbial pathogens or oxygen-demanding organic matter should leave the farm, contaminate the soil or influence the farm environment adversely.
- No excess amounts of odour should contaminate the surrounding air.
- No contaminated surface runoff or effluent should leave the farm.
- No discharge should give rise to 'material detrimental to any person'; that is, interfere with the normal use and enjoyment of life and property to an extent which is more than a trivial or minor nature.
- A feedpad should not be so aesthetically unacceptable as to reduce the value of surrounding properties.
- A feedpad should provide an environment that is conducive to the maintenance of animal health and the avoidance of animal stress.
- It should be constructed to withstand heavy traffic and be easily cleaned.
- It is recommended that a feedpad should be located at least 300 m from any neighbouring residence; this distance can increase depending on feedpad size and herd capacity.

- A landholder has a right to construct and manage a feedpad that meets regulatory requirements and accepted industry standards.
- Regulations governing feedpads may change over time requiring modifications during its operational life.
- Guidelines can only draw attention to the need to plan for change and the adoption of current best management practices for waste management, odour control and animal welfare. For example, future legislation may relate to greenhouse gas emissions and intensive land use.

Landholders proposing to install feedpads should look beyond the immediate reasons for constructing the pad and consider the likely long-term effects of their actions, and evaluate associated costs and benefits in detail. They must also consider existing and future neighbours and development prospects for their own and nearby properties. As is frequently stated, a dairy farmer is a custodian of the land and a steward of the environment.

A planning permit is generally required if the feedpad meets the definition of a cattle feedlot; the site is covered by an overlay (for salinity or flooding); the site proposal and works will affect flooding or discharge from the property; there is a proposal to remove natural vegetation; and/or a building is to be constructed. Catchment Management Authorities (in Victoria) may need to be involved in relation to potential impact on natural waterways. Rural Water Authorities may need to be consulted in relation to any potential impact on water supply channels or drainage schemes. Feedlot developments need to refer to relevant state codes for feedlot development.

It is recommended that neighbours are consulted to avoid any negative attitudes due to misunderstanding. Many agencies are influenced strongly by the number and strength of objections, emanating from neighbouring residences and communities.

The benefits of feedpads in grazing dairy systems

Integrating feedpads into grazing dairy systems has many potential benefits, such as:

1. Reduced feed wastage, allowing for better feeding of by-products and minimising high wastage rates, particularly when feeding out conserved forages.
2. Improved animal health, including more efficient rumen metabolism, a reduction in the incidence of mastitis, hoof and leg problems, better cow hygiene with less mud and manure and easier cow management.
3. Reduced climatic stresses because of the use of shade and sprinklers for heat stress, and wind breaks and roofs for cold stress.

4. Improved cow performance via higher feed intakes, increased control over diet quality, increased flexibility with feedstuffs, a higher milk yield, improved milk composition, greater feed conversion efficiency, improved reproductive performance, better growth in young stock, and better management of seasonal pasture feed gaps.

5. Protecting the environment via reduced degradation of paddocks and the direct use of effluent for recycling.

6. Better pasture utilisation via increased stocking rates hence grazing pressures, improved winter pasture growth (less damage), greater opportunities for conserving excess pasture for pad feeding, and increased flexibility to reduce grazing pressures during periods of poor pasture growth.

7. Higher farm profitability – smaller feed bills, the potential for reduced labour costs, an expansion of the dairy operation on existing farm area, better management of drought feeding, protection of farm infrastructure, all leading to the potential of an increase in cashflow and farm value.

8. Personal issues – at last, farmers can rest knowing their stock is safe. Peace of mind at night

Consequences of a poorly planned and designed feedpad

Unfortunately some farmers want to build a feedpad without giving it sufficient serious thought or even contemplating the potential consequences of getting it wrong. *Ad hoc* planning and poorly designed feedpads can provide unnecessary distractions and delays which snowball into a magnitude of problems. Mistakes in recent times have led to:

- *Financial burdens to rectify poor design.* Poorly planned feedpad siting, layout and design can have expensive ramifications. For example, poorly constructed foundations leading to ongoing repairs and maintenance, are all too common.

- *Animal welfare issues.* Insufficient animal spacing and poor layout can cause cattle injury and discomfort which in turn adversely affects cow performance. The spread of disease is also a risk that must be minimised.

- *Effluent discharges to nearby waterways leading to regulatory action.* The concentration of cattle in a confined area generates significant amounts of manure which require careful management. Nutrient runoff from feedpad areas, discharging off-site, along with odour emissions has the potential to attract unwanted regulatory attention.

- *Additional maintenance and repair costs.* With the constant movement of the herd on and off the feedpad and routine use of feeding machinery, wear and tear on the feedpad is a given. Unnecessary machinery and equipment breakdown becomes a burden unless rectified quickly.

- *Community protest.* It is common for new development to attract the attention of neighbours and government agencies. Ensuring proper planning and consultation is essential to dispel harmful rumours.

The role of feedpads in risk management

Farming is a risky business as any farm decision involves risk or uncertainty. If farmers knew what was going to happen, farming would be easy. It is preparing for what they don't know that is hard. Dairy farmers respond to risks by modifying their systems for producing milk. Farmers rarely depend entirely on grazed pasture because of the uncertainties of supplies of irrigation water, or rainfall in dryland regions.

Within southern Australia's dairy industry, farmers operate a diversity of farming systems, covering the entire range from spring calving herds fed entirely on perennial pastures through to year-round herds zero grazed and maintained on a feedlot. There is no single best system because individual farmers have, and are continually adapting production systems to meet the current farm-operating environment. As this changes every year, so to does their farming system. Feedpads provide the flexibility to modify their systems to meet the current production constraints. Gibb (2009) summarised the reasons why farmers choose to produce milk in various ways. Individuals have access to different resources, such as water, land, cows, capital, labour and management skills. They chose to use these resources in a range of different mixes according to their skills, experience and perception of risk and the constraints they find themselves under, such as debt load. Because conditions can change so quickly and unpredictably (such as the unprecedented milk price drop in late 2008), so to does their particular ideal best bet system. Often the lack of capacity to change, and the inflexibility to want to change can create unforeseen risks. Experienced farmers generally choose a system that they are comfortable with, taking into account resources, risk and constraints as they understand them. There are always costs in changing systems. These costs not only include direct cash outlays but also the time taken to learn to operate the new systems at high efficiency. Farmers then need to ask, 'How do I change my existing system to make it more profitable and less risky in the short and long term?'

Feedpads provide the flexibility to change farming systems relatively easily. Just because a farmer has a feedpad, doesn't necessarily mean he has to use it all the time. Many farms have machinery to make hay or silage but they only use it during spring when there is excess pasture and they need to build up their reserves of conserved forages for feeding out during the following autumn and winter or as a drought reserve. The same principle applies to feedpads. The relatively high cost of establishing feedpads usually means that many farmers feel they have to use them all the time. Feedpads are just like other pieces of farm machinery in that they should only be introduced to the short-term farm system when they are most needed to optimise profitability. With ample high quality pasture available per cow, it is generally more profitable to graze the herd and supplement with concentrates in the milking shed. As pasture supplies become more limited in availability per cow and/or forage quality, however, other supplements should be incorporated into the feeding program and this is when the feedpad can be used to aid feeding management and reduce feed wastage.

So establishing a feedpad is another risk management strategy; just one other way of increasing the flexibility to provide feed in addition to grazed pasture. As one farmer recently said, feedpads offer an insurance policy against poor seasons. The cost to the farm system will be discussed in Chapter 11.

The classification of feedpads

In trying to categorise feedpads, there are a wide variety of criteria available. Over the last decade or so, there have been at least five different classification schemes developed in Australia. The various teams of dairy specialists have differentiated between types of feedpads using many physical aspects, also using cost as a common denominator. As such classifications often overlap, these will be dealt with briefly below so a preferred classification can be developed for use throughout this book.

Published feedpad classification schemes

Davison and Andrews (1997) used cost per cow over 10-years old (so are somewhat meaningless), for three categories as follows:

- *Low cost* – cows are fed out on ground, along fence lines, roadways or using old tractor tyres or conveyer belts. These feedpads have high wastage rate and become very muddy in wet weather. Stabilised gravel can improve life of the feedpad.
- *Medium cost* – these usually include precast concrete troughs with concrete floor or cement-treated road base. They may or may not incorporate flood washing.
- *High cost* – these have concrete flooring and feeding area and are covered with a roof. They usually incorporate flood washing.

SIRIC (2002) based their classification on the material from which the feedpad is constructed, and its exposure to the elements which influence odour generation. Their classification is:

- *Dirt pad* – this is an open area either not formed or formed with a dirt pad. It has hay rings or minimal troughs.

- *Paved pad* – this is an open pad formed from impervious material with distinct feed troughs or feeding areas. The surface is a rock base, cement-stabilised clay with or without a geosynthetic material (thin flexible and permeable sheets of synthetic material used to stabilise soils).
- *Roofed pad* – this is a formed pad with ventilated roof and distinct roofed feed troughs or feeding area. The base is made of reinforced concrete or brick.
- *Enclosed pad* – this is a fully enclosed shed constructed of a formed concrete pad with a ventilated roof and distinct roofed feed troughs or feeding areas. The roof can be made of shade cloth (or mesh cloth), or iron and it can be a partial or complete roof over the feedpad. The lounging area can be an open lot or made of cubicles of free stalls

Kaiser *et al.* (2004) based their classification on the actual surface from which the stock eat:

- *Paddock feeding* – where chopped or baled silage is fed out directly onto the ground.
- *Bale feeder* – these are usually round but can also be linear feed barriers, and keep the stock off the feed to eliminate waste due to trampling and fouling.
- *Feed troughs* – these keep the feed off the ground thus preventing contamination by dust and mud. They also include feeding strips with a physical barrier to separate the stock from the feed.
- *Feedpads* – these are permanent feeding stations which provide cement surface for both the feeding strips and the cattle-feeding face.

Little (2007) and Dairy Australia (2007b) categorised four different feedpad systems based on their degree of permanency, capital cost, feed equipment used and expected feed wastage. These are summarised below and in Table 3.1 as follows:

Temporary and relocatable – these are low capital cost, quick to set up and have no prepared feedpad surface. The feed out facility can be relocated following rain.

Semi permanent – the feedpad surface is compacted and low cost troughs (conveyor belting, second-hand water troughs or other material from clearing sales) are used for feeding out. The feed out area can still be relocated to another site on the farm.

Permanent, basic but functional – this has a purpose-built feed out facility with a compacted surface, concrete troughs or a narrow cement strip and electric wires.

Permanent, minimal waste, maximum control – this purpose-built feed out facility most likely has concrete flooring and one or more feed alleys. It may be covered with a roof and may or may not incorporate a loafing area. The system would have a well-developed feed storage and mixing facilities.

Table 3.1 Features of different types of feedpads as described by Dairy Australia (2007b).

	Temporary and relocatable	Semi-permanent	Permanent and basic	Permanent and maximum control
Time and effort to set up	Very low	Low	Moderate	High
Weather durability	Low	Moderate	Moderate to high	High to very high
Permanency	Low	Moderate	High	Very high
Typical feeding equipment	Front-end loader or silage cart	Silage cart or mixer wagon	Usually mixer wagon	Mixer wagon
*Capital cost ($/cow)	<$50	$50–100	$100–200	>$200
Feed wastage and (% loss)	Very high (>30%)	Moderate to high (15–30%)	Moderate (8–15%)	Low (5–8%)

* Cost is only for the feed out area, excluding associated feeding equipment.

McDonald *et al.* (2008) classified feedpads based on the stage of farm development, which obviously includes capital investment, summarised below and in Table 3.2 as follows:

- **Designated sacrifice area** with scattered hay rings and simple fencing to create several pens.
- **Formed earth pad** – this can be made from compacted earth, rock base or compacted clay.
- **Introduction of troughs** which allows feeding of formulated supplements via specialised feed out wagons, but necessitates installing water troughs and collection of nutrient runoff. This required pad-scraping machinery and effluent-solid stores.
- **Concreted sloped pad** with nib walls thus allowing for washing systems, hence requiring sump collection effluent storage and potential reuse. This stage of development is usually associated with feed bunkers, mixer wagons and even the installation of head stalls.
- **Free stall shed** – this includes the installation of a canopy or roof and eventual stalls and bedding and a large loafing area. Such capital investment also includes advanced solid separation systems, specialised machinery (such as bobcats), rainwater diversions and if required, cooling fans and sprinklers.

Preferred classification system for feedpads

Each of the five classification systems described above have their merits, but there is considerable overlap and inconsistencies in terminology. For example, Kaiser *et al.* (2004) uses the term 'feedpad' for the most sophisticated feedpad system. In the SIRIC (2002) system, the two most sophisticated systems (roofed pad and enclosed

Table 3.2 Pros and cons of different feedpad options as described by McDonald *et al.* (2008).

Feedpad type	Pros	Cons
Designated sacrifice area	Designated feeding site close to the dairy shed	Poor draining causing pugging
	Opportunity to relocate to other farm area as needed	Extensive wear and tar around high traffic areas
	Low cost to establish	Restricted to hay or silage supplements
	Specialised equipment not required as front-end loader adequate	Potential for environmental mastitis problems
	Easy to remove hay rings for cleaning	Significant wastage of supplements
	Dry scraping sufficient	Continuous damage to hay rings from machinery and cattle
	Opportunity to utilise batch feeding	Limited use during wet weather
	Various modular layouts available	Cow bullying due to limited feeding space
Formed earthen pad	Solid stable foundations	Potential for surface water pooling
	Lower maintenance	Necessity to constantly dry scrape
	Low cost to install	Potential to become slippery for cows and machinery
	Gentle slope to diver rainfall runoff	Feed wastage
	Capacity to handle larger herds	Potential herd health issues
Formed earthen pad troughs	Contains feed and reduces spillage	Feed spoils may get trapped underneath creating odour, which hinders feeding
	Relatively cheap to set up	Potential for cattle injury or death as cows can fall in troughs
	Layout can be doubled up to suit the designated area	Difficulties in cleaning in and under troughs
	Opportunity to feed a wider range of supplements	Trough height may not be compatible with feed out machinery
	Opportunity to fence section off to cater for other herds	Tendency to hold water during rain events damaging feed
Concrete pad	Solid, stable foundations capable of handling large herds and machinery	Usually require large volumes of water washing
	Cleaner environment for cows to spend longer periods	Extended period may cause feet problems
	Feed wastage significantly reduced	Alleyways may require regrooving due to constant wear and slippage
	Opportunity to reuse recycled effluent water	Cows may push feed out of reach

Table 3.2 continued

Feedpad type	Pros	Cons
	Multi-purpose facility to cater for needs of different herds	Increased effluent loading requiring management
	Permanent high value farm asset	Contingency plans required when pumps fail
Free stall shed	All-weather facility to control herd comfort	Significant capital investment
	Reduces costs associated with managing pastures	Extreme water usage in cleaning requires a recycling option or large water right
	Reduced farm capital infrastructure such as laneways, fencing and stock troughs	Potential for other environment issues such as noise and odour
	Opportunity to monitor the herd closely	Labour-intensive
	Opportunity to increase feed imports to match seasonal variations	Difficulties in retaining bedding material from the effluent stream
	Minimal feed wastage	Advanced effluent system needed with constant maintenance
	Minimum cow travel sustaining production	Difficulties accessing sick or injured stock
	Clean and dry environment for herd	Specialist machinery and equipment in cleaning and feeding
	Opportunity to manage large herd on smaller enterprises	Necessity to deal with many agencies and authorities

pad) incorporate some form of overhead cover whereas such a structure would only be expected in the most sophisticated systems described by both Dairy Australia (2007b) and McDonald *et al.* (2008), being the 'Permanent and maximum control' and 'Free stall shed' respectively.

For an all-encompassing system of classification, several criteria should be selected to differentiate different types of feedpad systems more clearly (also using the analogy of different rooms in a house as discussed in Chapter 1). These are described in Table 3.3. There are three options for feedpad surface for the stock (the dining room); Option 1 provides six choices of feeding system (the dining table), and Option 2 provides four choices; three options for the lounging area (the lounge room); and three options for overhead cover for feed and/or stock (the roof).

Additional criteria could be included in a classification scheme as useful descriptors, but these do not necessarily add extra information about the feedpad area itself. These include three options for the form of effluent management; four options for the machinery for feed preparation and delivery; and three options for the type of feed storage.

Table 3.3 lists the classification of each criterion. Using four major criteria would lead to a very complex classification system, so some shortcuts are desirable. These are the types of shortcuts used in the previous classification schemes. For example, the low-cost feed troughs could be combined with the low-cost feed strip and the concrete troughs and concrete feed strip could be combined as a concrete

Table 3.3 Suggested criteria used in a feedpad classification scheme. There are two options provided for feeding system to reduce the complexity of the classification.

Classification criteria:	Options for each criteria
A. Feedpad surface	1. Dirt or a grazed paddock 2. Formed earth; also includes formed gravel, rock and clay with or without geosynthetic sheets 3. Cement
B. Feeding system (Option 1)	1. Paddock; this could be a grazed paddock or a sacrifice paddock 2. Hay rings in paddock 3. Low-cost moveable troughs, also includes conveyor belt, linear feed barriers 4. Low-cost feed strip, such as formed earth, rock and clay 5. Concrete troughs 6. Concrete feed strip
B. Feeding system (Option 2)	1. Paddock 2. Hay rings in paddock 3. Low-cost moveable troughs and feed strips 4. Concrete troughs and feed strips
C. Lounging area	1. Sacrifice paddock 2. Loose housing on lounging pad or open lot 3. Free stalls
D. Roofing	1. None 2. Shade cloth 3. Permanent
Additional criteria:	
E. Effluent management	1. None 2. Dry scraping 3. Water: hoses 4. Water: flood washing
F. Feed preparation & delivery	1. Front-end bucket 2. 'Side winder' round bale feeder 3. Silage cart 4. Mixer wagon
G. Feed storage	1. Silage storage 2. Open commodity bunkers 3. Covered commodity bunkers

feed out system. This would reduce the Feeding system options to four, as in Option 2 for Criteria B in Table 3.3.

The two major classification criteria would be the feedpad surface and feeding system and these would form the major basis for describing individual feedpads. The lounging area and overhead cover are useful descriptors but would be in part, self-selected by decisions made on the feedpad surface and feeding system. For example, you would only expect free stalls and a permanent roof in a feedpad with a cement surface and a concrete feed out system.

What is the maximum number of categories for a useable scheme is debatable, but any more than six or eight may make it too complicated. The three additional criteria in Table 3.3 are certainly useful in providing extra information about any feedpad system. A fourth additional criterion could be the proposed maximum (or average) daily hours of usage of the feedpad, but this is likely to vary widely in different seasons. With regard to the most up-to-date and accurate estimates of installation costs and rates of feed wastage, the Dairy Australia (2007b) scheme, outlined in Table 3.3, would be the most relevant to use.

Stages of feedpad development using farm plans

Many feedpads evolve over time, starting as a simple feeding facility, eventually developing into a sophisticated concrete construction incorporating a lounging pad as well as feeding area. The development of farm plans can help with identifying issues, setting priorities and planning how to address them. They can prioritise the work, be used to formulate budgets, stage development and then track and improve farm performance while planning for unexpected events (Dairy Gains 2008). A number of different plans can be used to address specific issues and to assist with planning applications as well as ongoing site management, ensuring all staff understand the operational objectives. Typical plans may include a *Contingency Plan,* for procedures and contact details for unexpected events, such as machinery breakdowns and disease outbreaks; a *Whole Farm Plan*, to focus on the overall management and layout of the entire farm including improvements and infrastructure; an *Environmental Management Plan* to focus on general management of the farm, taking into account the environment and associated risks; an *Effluent Management Plan*, that is the technical design and management of effluent, focussing on effective use of nutrients; a *Nutrient Management Plan,* for nutrient budgeting and mapping, focussing on the productive use of nutrients across the farm; and finally, a *Feedpad Management Plan,* to ensure the effective integration of the proposed feedpad into the whole farm system.

McDonald *et al.* (2008) have suggested a series of trigger points, which require the consecutive development of farm plans. The various scenarios can evolve as follows:

1. Low-use earthen feedpad for supplementary feeding, which requires a Feedpad Management Plan.
2. Earthen pad area with more than 500 m^3 of rainfall runoff or the introduction a wash down system, which requires an Effluent Management Plan.
3. Cattle confined to an area containing 50 dairy cattle units (see Chapter 4) or more, which requires a Nutrient Management Plan.
4. Significant changes in feeding, with more of the feed imported onto the farm, which requires a Nutrition or Supplementary feed budget.
5. Housing the stock for extended periods may require an environmental risk assessment and assessment of animal welfare.

When starting to prepare the Feedpad Management Plan, it is important to be aware of other potential impacts such development may pose on the farm. The initial approach may simply be to develop a Feedpad Management Plan; however, there is a likelihood, depending on the scale and stage of development, that it may trigger additional plans. For example, an Effluent or Nutrient Management Plan would be necessary once the development entered the correct stage or the proposed cattle occupation time were to increase significantly. Feeding strategies and farm water budgets also need to be reviewed and aligned with the new changes to enable the farm to maintain its productive performance.

These plans must allow for expansion and modification of existing systems with long-term changes in overall farm management. They should be tailored to suit the unique characteristic of each farm. In addition, they should be developed in collaboration with suitable trained service providers and regularly reviewed and adjusted to reflect current and future farm practices.

Until quite recently, there were no dairy industry codes of practice or guidelines addressing the development of feedpads or free stall sheds. The guidelines developed by SIRIC (2002) and O'Keefe *et al.* (2010) introduced a broad framework of recommendations for the planning, design and management of cattle feedpads in Victoria for use by planning authorities and design consultants. They include detailed information regarding statutory planning, engineering design and maintenance, animal health and welfare, manure management, supplemental storage and occupational health and safety. Their aim is to help councils and design consultants to determine whether a planning permit is required and provide technical information for preparing and assessing planning permit applications. They are based on two key statutory outcomes, namely environmental protection and animal welfare.

It should be noted that these outcomes are not the drivers of this manual. Obviously, farmers contemplating feedpad development must be aware of these statutory outcomes, but this manual has been written to provide an overview, in

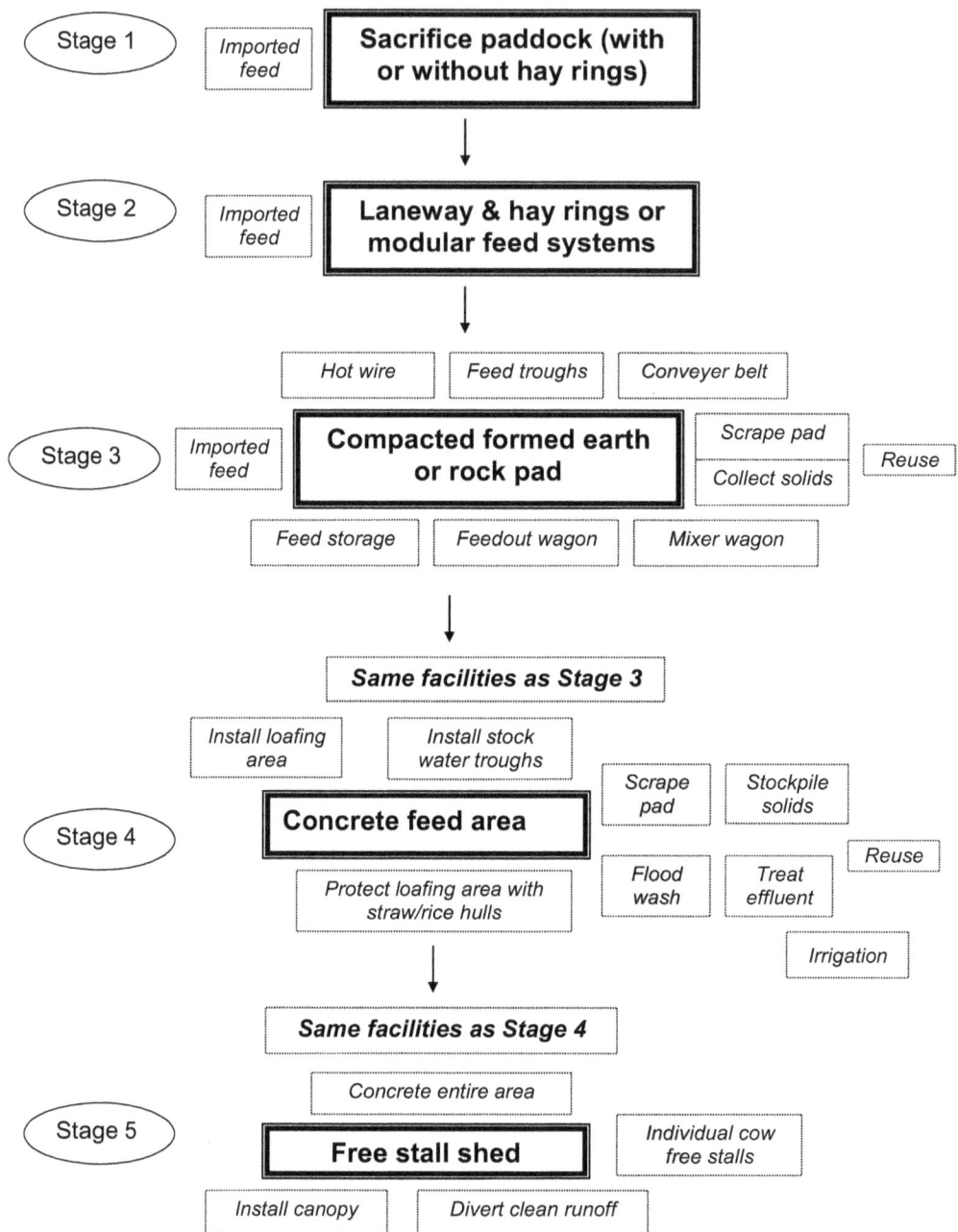

Figure 3.2 The five stages of on-farm development of dairy feedpads.

some cases quite detailed, on all the on-farm implications that dairy farmers should consider when planning to intensify their production system by investing in more efficient feeding facilities. Although the emphasis is on producing more milk from the same amount of feed, or even the same quantity of milk from less feed, any change in feeding management invariably influences other aspects of herd and farm management.

The on-farm development of feedpad technology

Only when farmers become more dependent on increased amounts of imported forages does feedpad technology become important. During their twice-daily milking sessions, grazing cows can physically consume more than 10 kg/cow/d of concentrates, supplying most if not all of their high energy feed requirements. Consequently, the milking shed can be the sole dining room for concentrate feeding of grazing cows. Because the simultaneous feeding of additional forages will optimise rumen function (see Chapter 7), however, the pulse feeding of concentrates over two 8–10 minute periods each day is not the most nutritionally efficient method to provide these energy-rich feeds which are necessary to ensure high milk yields. Furthermore, feeding forages is more difficult than feeding concentrates because they are generally lower in nutritive value hence more bulky and invariably lead to greater wastage rates. Clearly feedpad technology provides a more effective way to feed out supplemental concentrates and forages.

Installing a feedpad usually comes about in a natural progression and development of the farm. For example, a farmer may begin by feeding hay out in the paddock during winter, then progress to feeding hay in a laneway at other times during the year. He may then develop to a more substantial raised pad for supplementary feeding, prior to deciding to install troughs and a roof, and then finally construct a free stall shed.

Figure 3.2 provides a flowchart of the five stages of on-farm development of dairy feedpads, with the associated facilities as the system becomes more sophisticated. A series of photographs are presented for each stage in Colour plate 1.

Colour plate 1   The five stages of on farm development of feedpads: (a) sacrifice paddock with hay rings, (b) modular feed systems, (c) compacted rock and feed trough, (d) concrete feeding area, (e) free stall shed.

Colour plate 2   The foundations of laneways require close attention to ensure they are weather-proof.

Colour plate 3 Laying the foundations for a loafing area.

4

Physical aspects of feedpad design

This chapter discusses the physical features of feedpad design and construction and provides a checklist when planning a feedpad.

The main points in this chapter:

- The capacity of the system depends on the number of stock to be fed, their average live weight and the length of time each day they spend on the feedpad.
- When siting the feedpad, considerations should be given to the topography, soil type, potential impacts on ground and surface water and plans for effluent management.
- Feedpad design depends primarily on land slope and surface construction material, which can range from formed earth to gravel to concrete.
- The size and layout varies with the planned capacity of the feedpad, the drainage and the inclusion of loafing pads or free stalls.
- Each laneway should have a sound base to be long lasting. Yard surfaces must not be too slippery or abrasive.
- Other aspects of feedpad planning should include minimising odours and dust and also potential noise. Landscaping can improve the aesthetics and reduce negative perceptions of feedpad operations.
- Stand off pads, specifically to reduce pasture damage during wet weather, are effectively loafing pads associated with feedpads.
- To aid planning, this chapter provides a checklist of all the above considerations.

Choosing the most appropriate feedpad system to suit the farmer's objective and management strategies is essential to ensure the proposed development enhances the long-term viability of the farm. The feedpad system, irrespective of scale or whether it is a temporary relocatable hay ring system or more permanent concrete pad, requires detailed planning, taking into consideration the farm's individual

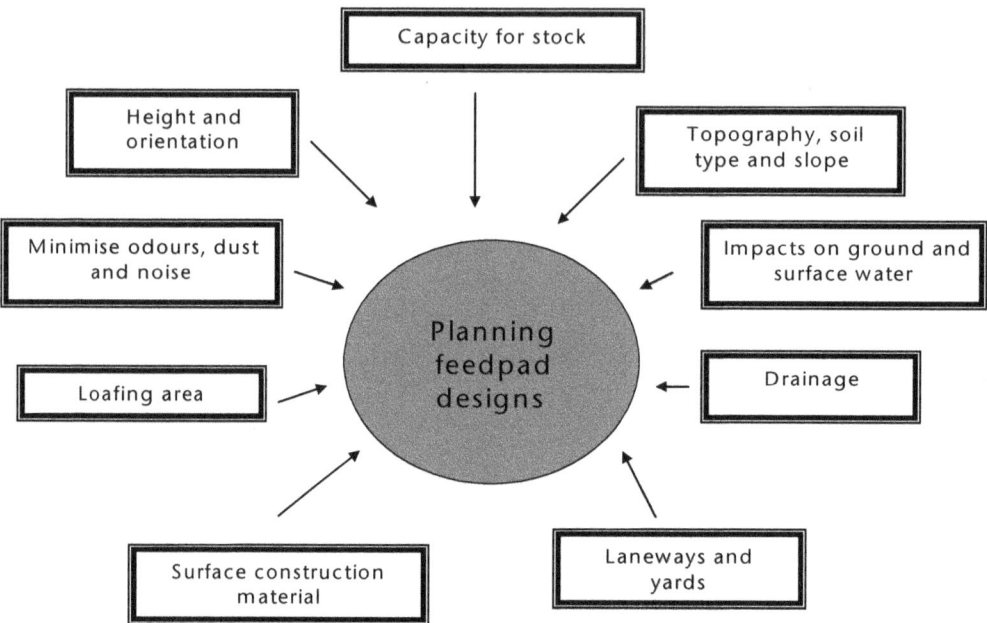

Figure 4.1 The key considerations when designing a feedpad.

characteristics and locality. For the feedpad system to be effective, it must be carefully designed, sited and integrated with the farm's existing or proposed infrastructure (see Figure 4.1).

It is paramount in the design stages to address potential community concerns by considering potential adverse impacts to sensitive areas such as neighbouring residences and nearby waterways. In addition, issues such as animal comfort and welfare need to be addressed.

Capacity of feedpads

To develop consistency and a general basis for comparing feedpads, the *Dairy Cattle Feedpad Guidelines for the Goulburn Broken Catchment* (SIRIC 2002) have utilised the concept of 'Dairy Cattle Units' (DCU) which are similar to the Standard Cattle Unit developed for use in legislation of beef feedlots.

This calculation removes the myth that larger herds pose the bigger risk to the environment by taking into account potential manure loading and occupation of a specific area. The DCU is therefore an important calculation in setting appropriate buffer distances from sensitive areas.

The DCU is based on the number of dairy cows on the feedpad; their average live weight, using 550 kg as 1.00 units; and the duration of each day they spend on the feedpad.

Table 4.1 Conversion factors used to convert average live weight when calculating Dairy Cattle Units.

Live weight (kg)	Conversion factor
300	0.70
350	0.76
400	0.82
450	0.88
500	0.94
550	1.00
600	1.06
650	1.12
700	1.18

(Source: SIRIC 2002)

Use of a standard weight allows for adjustments for stock with different average live weights as in Table 4.1.

The number of hours stock spend on a feedpad will vary with stage of lactation, pasture availability and season. For planning purposes it is important to adopt peak loading conditions based on maximum duration cows spend on the feedpad. This may not account for exceptional circumstances such as very heavy rain which would limit access to grazed pasture. Monthly averages for daily durations can be used in peak loading calculations. Thus for a herd of 400 cows weighing on average 600 kg and spending an average of six hours per day on the feedpad, the calculation is as follows:

$$400 \times 1.06 \times (6/24 \text{ or } 0.25) = 106 \text{ DCU}$$

Siting the feedpad

Feedpads need to be sited to ensure a balance between dairy production, livestock health, environmental protection and long-term farm sustainability, and at the same time, having minimal impact on neighbours. The first step is to identify all farm assets and infrastructure relevant to the farm that may need to be protected from an operating feedpad which generates high nutrient loads, as well as existing farm infrastructure that may need to be incorporated into the feedpad operation.

There are many issues that must be considered when selecting a feedpad site. These include access to the milking shed, feed supplies and farm laneways for stick movement; access for employees and vehicle; expansion opportunities and staged development plans; access to sufficient water of acceptable quality; slopes and other topographical features. Consider also visibility from the milking shed and farm residences as well as any existing site services such as channels, drains and electricity; existing vegetation (particularly native); proximity to waterways, dams and streams, and existing buildings (particularly farm residences). Think about the feedpad's prominence in the landscape as well as views from the site, general aesthetics and amenities; the provision of shade in the absence of trees on the farm; boundaries and easements; flooding impacts, groundwater impacts and consequences; and uncontaminated stormwater control or diversion. How easy is it going to be to harvest the manure and reuse wastewater and nutrients? How will the prevailing winds affect the farm and its neighbours, particularly in relation to odours? Might there be potential conflicts looming with regard to future development of urban areas, residences and land zoning? Are there any archaeological or heritage sites nearby that might be affected? Any threatened or endangered species or ecological communities? And how close are services such as underground telecommunications, water reticulation, and electrical and gas transmission conduits?

Other industry guidelines for effluent storage and feedpads require the following buffer distances:

- 1 m above the highest seasonal watertable, especially for the base of effluent ponds, feedpads and manure storage areas
- 45 m from the dairy milk vat room
- 50 m from closest property boundary
- 60 m from infrastructures built by Rural Water Authorities, such as channel and drainage networks
- 200 m from water bores
- 200 m from flood-prone land
- 300 m from neighbouring residences
- 800 m from any potable water supply take-off controlled by a statutory authority.

Many State Environment Protection Agencies do not have set buffer distances and use those above as guidelines. The objective is to show clearly that the proposal will have no adverse impact on sensitive areas in close proximity to the development. A key step in feedpad siting is often determined by working on a process of elimination of where it cannot be located.

Topography

The natural drainage regime of the area must be considered to facilitate the diversion of uncontaminated storm runoff, the removal and transfer of wastewater to an effluent storage, the reuse of wastewater (particularly if planning to use an existing farm reuse system) and the drainage of contaminated runoff from the feed storage area.

Slope creation on a feedpad relies on having sufficient soil depth to accommodate the cut, fill and borrowing requirements necessary for undertaking earthworks. This applies particularly to areas where holding ponds are required.

The pad should be formed well above the natural surface level to promote drainage, increase air movement to decrease odours and to discourage insects. The cost of sourcing and moving sufficient suitable soil need to be taken into account. Buildings and works should not be located on steep slopes greater than 20%.

Flooding

Before determining a potential feedpad site, a property floodplain overlay should be obtained from relevant local or state government agencies to clearly identify potential flood prone areas. This should avoid any adverse impacts during significant rainfall events and periods of prolonged rainfall.

The feedpad should be located above the one-in-100-year flood level and all earthworks should be designed to avoid off-site impact of floodwater discharge either through funnelling or backwater effects. The tops of any farm turkey nest dams should also be above this flood level, but not necessarily any irrigation recycling sumps used to shandy effluent.

Groundwater impacts

Groundwater investigations are also important when designing feedpads and determining potential sites for construction. It is often advisable to dig check wells at various localities and monitor them over a 24 hr period to see if they fill with water.

Feedpads and associated areas such as cattle loafing and silage bunkers have the potential to generate high levels of nutrients which in turn pose significant risks to both surface and groundwater supplies through surface runoff and leaching. These risks need to be identified and managed during the development phases of the feedpad.

Pad slope

A fundamental design aspect of any feedpad or free stall is ensuring the pad slope is adequately determined and constructed to optimise the proposed site's natural topography and to avoid unnecessary soil importation which will add significant cost to the construction. O'Keefe *et al.* (2010) provide detailed design guidelines relating to slope and therefore should be consulted before undertaking any construction.

The overall slope of the pad and associated loafing areas is important to remove effluent, fibrous material and sand from alleyways by optimising flood-washing systems performance; enable sufficient drainage from the site to remove water quickly following heavy rainfall; prevent unnecessary pooling of water which creates odour and boggy areas; enhance the flow of cattle and precent flat, slippery areas; utilise the site's topography and natural drainage features; and eliminate tractor and mixer-wagon slippage when feeding out.

Soils

Soils need to be evaluated to assess their suitability for constructing the pad and effluent storages. The engineered earth must be strong enough to carry the heavy loads of machinery, building foundations and of course, stock (on an earthen pad). Clay-dominant soils will be required and sandy soils avoided. Soil investigations and permeability tests should be undertaken to avoid costly mistakes and risk to the surrounding environments.

During feedpad construction, better compaction can be achieved by removing and stockpiling the topsoil for later reuse for landscaping or covering exposed soils (with a minimum layer of 100 mm to protect against erosion).

Insufficient soil investigations for feedpad sites can lead to constant maintenance and repairs of high-traffic areas; pad surface cracking; groundwater pollution from nutrient leaching; and seepage into effluent storages.

Effluent and manure management

Irrespective of the scale of development of any feedpad, the design of an appropriate effluent system and the management of both liquid and solid manure is a major consideration that requires thorough planning. Effluent system design is a specialist field of expertise and therefore the need to consult a professional is essential. Further details are provided in Chapter 5.

Feedpad design

Feedpad type

The type of feedpad will depend on many aspects such as the proposed purpose of the feedpad; herd duration and occupation periods on the area; financial resources available; suitability of the proposed site; the feed handling system and conveyance method to feed stock; the type of feed storage (vertical or bunker bins or silos); the type of feed being fed out (grain, by-products, silage and/or fresh forage); mode of cleaning the pad and methods of handling manure; and type of machinery and equipment being used.

Surface construction material

The feedpad surface should be designed to provide sufficient slope for maintenance of gravity drainage; prevent surface wastewater reaching the subsoil; have a deep and stable foundation to apply loads to the ground without causing settlement; minimise stress, disease and injury to stock; provide a durable, clean working area of the required profile and a non-slip finish; facilitate cleaning without surface material being removed. At its simplest, it should contain a compacted interfacial layer of manure and soil to form a biological seal to decrease water infiltration into the underlying soil. Low infiltration rates restrict leaching of nitrates, salts and ammonium into the subsoil, thus protecting the groundwater resources of the site from contamination.

Ideally the pad should be evenly graded and compacted to form a smooth impervious surface. Materials used for pad construction are many and varied and will govern construction costs and pad longevity. The type of surface influences stock health, particularly mastitis, traumatic injuries and hoof well-being. Low infiltration is important to minimise leaching of nutrients, pathogens and salts into the subsoil. Feedpad construction can be based on a series of surfaces such as earth, to stabilised earth and eventually to concrete. Considerations for each surface include:

- *Earth* – compaction is important to reduce permeability of soils on feedpads or effluent storages. Over time, an earthen pad usually develops a compacted layer of manure and soil and this can form a biological seal that decreases infiltration of water. It may be worthwhile to incorporate chemical stabilisers, such as hydrated lime or gypsum to maintain integrity of clay surfaces. Sub-surface drainage can also be incorporated using lines of slotted drainage pipes, 1.5–2 m apart, overlaid by 20 cm gravel. If desired, these surfaces can then be covered with rice hulls or sawdust.
- *Gravel and coarse sand* – it is important to select carefully graded material to aid compaction of the pad surface, avoiding material containing sharp stones.
- *Geosynthetics* – these are thin, flexible and permeable sheets of synthetic material used to stabilise and improve the performance of soil in civil engineering works. They are cheap, resistant to soil chemicals, moisture and bacteria, and used for filtration which restricts movement of fine soil particles while allowing the soil to remain permeable to water movement. They also reinforce and stabilise the soil to decrease soil compaction by stock. As water is conveyed vertically or horizontally, drainage is aided along the material to any effluent outlet.
- Materials like **bitumen, bricks and concrete blocks** provide the most durable surfaces but are the most costly
- *Concrete* – it is important that these surfaces do not become slippery. Avoid this by scoring or grooving during construction, such as using heavy gauge mesh over the wet concrete. The finish must be rough enough so cows will not slip while still able to be cleaned and not cause hoof damage during long-term occupancy.

- A permanent barrier such as a nib wall and posts separates the drive and feed alleys. This 'cow barrier' ensures unrestricted access to the drive alley without disturbing the cattle on the feed alley. The drive alley can be constructed with an inverted triangle profile to shed runoff from the feeding table. Runoff from the pad should be contained and managed.

Specifications for concrete

The actual concrete strength and thickness recommended depends on the type of strength of the sub-base and base materials and the proposed animal loadings. O'Keefe *et al.* (2010) have listed the specifications for concrete in feedpads as:

- Footings for foundations: N20 concrete.
- Pedestrian traffic: N25 concrete.
- Cow and vehicle traffic: N32 concrete.
- Drive alleys: 150 mm thick with SL82 reinforcing mesh in the top layer.
- Feed and cow alleys: 100–125 mm thick with SL82 reinforcing mesh in the top layer.

Alley designs should incorporate construction, expansion and contraction joints, placed in regular grids and no more than 4 m apart. These should be designed in advance, placed in the overall layout and be designed to only move in horizontal directions.

Recommendations for steel and post, wire and cable specifications and recommended slopes of feedpad surfaces are provided by O'Keefe *et al.* (2010).

Feedpad sizing and layout

The size of the feedpad can be determined by the farm management as to whether the feedpad will be developed to accommodate the entire herd or in feeding batches. The feedpad should be designed for a logical layout of adequate facilities aimed for efficient operation, good stock health and environmental protection. Such designs should also allow for herd expansion in the future. Space requirements for stock on a feedpad are usually a function of how the pad is to be used and how long the stock will stay on it.

Enough physical space and separation/buffer distances should be provided for future expansion or staged development. Examples could include increasing herd size, requiring extension of feedpad and effluent management facilities; incorporating partial or total mixed rations into the feeding system, requiring additional space for commodity bunkers and feed preparation areas; converting permanent concrete feedpad to free stall shed, necessitating additional space for constructing free stall cubicles; and changing feedpad cleaning system from scraping to flood washing, necessitating additional space for water tanks, containment sumps and ponds to manage feedpad effluent.

Access for stock and vehicles

The feedpad should be in a central location to minimise vehicle movement and allow for orderly stock management and effluent collection.

The physical dimensions of the laneways, races, gates, entrances and exits should be designed for ease of stock access and movement. Vehicles require a minimum of 3.7 m for easy access. A compromise is usually necessary between very wide laneways, which occupy land and reduce stock control and narrow laneways with a funnelling effect which can contribute to problems with lameness. Structures should take advantage of the social behaviour and natural movement of cows; fences and gate should have no protrusions to hinder stock movement. Laneways should be a minimum of 4 m wide which should be increased to 5.5 m for herds of more than 120 cows. Laneways 8–10 m wide are now commonly used on farms with herds in excess of 1000 cows. Very wide laneways can reduce control of stock movement. Allow stock to enter the feedpad without having to change direction by more than 90°. All laneways, access and stocked areas should be well drained with the runoff directed to drains leading to the feedpad effluent system.

The feedpad design should also allow for all weather access of vehicles with minimum width for laneways of 4 m. Feed lanes should be 4.5 to 6 m wide to allow for easy tractor and feed out wagon access. Entry and exit points, as well as turning areas for cleaning and feeding out, should be wide enough (8–10 m) to allow free flow of stock and vehicles. There must be sufficient room for distribution of feed on the feedpad as well as room for cleaning operations; the easy movement of trucks to deliver feed on site; access from a main or secondary road; sufficient room for large trucks requiring high clearance; and adequate area for collection and stockpiling of solid wastes and residues from effluents storages and solids traps.

Constructing and maintaining laneways

Routes for laneways should permit easy cow flow and, if necessarily, allow for herd expansion. Grass and topsoil should be removed and the prepared area well compacted, above the water table and free from pugging (see Colour plate 2). Additional materials such as pit or river gravel may be necessary to elevate or stabilise the surface in preparation for track construction. This layer should also be well compacted.

Firstly, construct a sound base that can support the top surface without moving, wetting or breaking up:

- Using coarse material with a clay content of 15–30%. Do not use soft clay.

- Applying the material in 150 mm layers and compacting well between layers with a roller at slow speed. Use a power grader with an experienced driver, or alternatively, a tractor-mounted blade and a pneumatic -tyred roller or loaded vehicle with high-pressure tyres.
- Developing a camber or slope across the track that will shed water and be comfortable for cows, using a slope of, say, one-in-10.

Secondly, construct the top (wearing) surface suitable for cows to walk on, ensuring it has a minimum depth of 50 mm, with 100–150 mm preferred, and that it does not contain stones which can cut hooves, or free uneven material that can cause bruising. If coarse gravel is used, it is well-rounded gravel of less than 25 mm diameter, with 15–30% clay content to allow adequate compaction, binding, wear resistance and smoothness. Large stones should be avoided as these get kicked out of the track, leaving the site susceptible to water ponding and hoof damage. The surface should be sealed and the camber finished to prevent moisture draining into the base. The addition of 3% cement prior to compaction assists sealing. Compaction is essential as cows are not able to compact tracks adequately.

When maintaining laneways, seek out sections that are breaking up to identify and rectify any underlying causes. Drainage may need to be improved or the top surface requires additional compaction. Repair sections closest to the dairy first, as these are ones most frequently used. Remove stones from these sections and make sure they are well compacted, appropriately crowned and well drained. Cows cannot compact these wearing surfaces adequately. If problems exist with cow flow, that section of the track may need drainage, repairing, crowning, maybe widening or even rerouting.

It is important to avoid right-angle bends in laneways as cows tend to bunch up, slowing down movement and increasing effluent dropped. Manure in laneways is less likely to enter the effluent system to be redistributed around the farm, thus losing its potential as a useful soil nutrient. Good cow flow minimises the time cows spend on laneways, hence the lost effluent. Allowing cows to move at their own pace will also reduce fouling of laneways.

Entrance to the feedpad

When cows step from a gravel race to a concrete surface, the small stones caught in their hooves are a major cause of lameness. To avoid this, there are three main options available:

1. Use a softer surface leading to the entrance. This can be limestone, soft rocks or sand spread 150–200 mm deep over the last 100 m leading to the concrete.
2. Install a footbath system to wash away the stones. It can also contain copper sulphate or zinc sulphate solution to improve foot health.

3. Construct a step barrier so cows have to lift their legs to step over it thus allowing stones to drop out of the hooves. These also act as nib walls to contain feedpad effluent.

Yard requirements

The space required by a 600 kg dairy cow varies from 200 m²/cow for a grazing animal to 9 m²/cow for a cow spending 24 hr/d on a feedpad to 1.2 m²/cow standing room in the yard before milking. All cows will generally stand if only on a feedpad for a few hours. Up to 10% will lie down if constrained for 5–7 hr, however, while all will want to lie down over a 10 hr period. Cows need to lie down for at least 8hr/day and if restricted, they will prefer this to grazing, which can lead to underfeeding.

The type of surface influences the desire for cows to lie down. Cows will lie down sooner on softer surfaces compared to harder surfaces. They are less likely to lie down on wet or very slippery surfaces.

On a feedpad, cows should then be provided with:

- 3.5 m²/cow when used for short periods.
- 6 m²/cow when used for longer periods, say 12 hr/day.
- 9 m²/cow when used permanently with no grazing, plus an additional 1 m²/cow feeding area.

When constructing yards, discuss the surface preparation with your contractor, making sure hoof wear is considered when pouring the yard surface. The surface should not be too slippery or abrasive.

The surface must be kept clean by removing sand, gravel and stones every time it is washed. Broken concrete should be removed and the area repaired. If sole abrasion is identified as a cause of lameness, check the yard surface, taking into account the length of time stock have to stand on it. If cows slip frequently on the concrete, clean it with a high-pressure hose or consider grooving the surface.

A set of working yards (including races, crush and even a drafting gate) should be located close to the feedpad. This should include lights for emergency night visits from veterinarians, with the crush under cover for all-weather use. Sufficient space should also be left for a set of cattle scales in the future. An office and staff amenities room can service the needs of the milking shed as well as the feedpad complex.

Height and orientation

The height of any shed should be accessible to machinery used for cleaning and feeding, with a minimum recommended ridge height being 7 m. Removable roof panels provide flexibility. It should also promote sufficient ventilation and penetration of sunlight. As the positioning of eaves and roof canopies will affect shading and sunlight, orientation is important. Winter sunlight provides additional warmth for the stock, helps dry the feedpad surface and reduces the incidence of diseases. Summer shade reduces heat stress.

The feedpad can be orientated to either maximise or minimise exposure to climatic elements such as prevailing winds, sunlight and rainfall. Extreme weather conditions can affect all feedpad operations from time to time, regardless of geographic orientation. Orienting the feedpad perpendicular to the summer and autumn prevailing winds assists in cooling the cattle. For a covered or shaded feedpad, a north–south orientation promotes drying because the shade moves across the pad during the day. Dry pad conditions are important as wet areas can predispose the stock to mastitis. Since a north–south roof or shade orientation does not provide any less area under shade compared to an east-west one, covered free stall sheds should be orientated east–west.

The recommended orientation tends to vary from one part of Australia to another. A north–south alignment is favoured in Queensland whereas in Victoria, an east–west alignment is popular. The major reason is that east–west orientations maximise shade whereas north–south orientations maximise sun, to promote drying and reduce bacterial populations.

Drainage

Effective drainage is important for all-weather access, collection of contaminated runoff and diversion of uncontaminated stormwater. Raised feedpads will promote natural drainage. The drainage system should be designed to handle feedpad runoff from a one-in-20 year 24-hour storm event. Additional water from all surfaces, as well as washing systems and trough spillages, should be included in any storage volume calculations.

The drainage system should incorporate drains or diversion banks, a sedimentation basin to remove solids from liquid effluent, and catch drains to carry storm runoff and effluent, with minimum slope of 0.5%.

It is important to minimise the necessary capacity of effluent storage requirements. Rainfall contaminated by faeces, urine or residue feed should be directed into effluent systems. Uncontaminated rainfall runoff (from clean surfaces such as rooves) should be diverted to the natural drainage system.

Runoff from livestock areas contains relatively high levels of nutrients, salts, chemicals, debris, pathogens and oxygen-demanding organic matter so must be directed to effluent reuse systems. Wherever changes are made to cow flow, provision should be made for the diversion of contaminated runoff to effluent systems. Further details are provided in Chapter 5.

Loafing pads

These provide area for cows to relax following feeding and can be equipped with shade (and fans plus sprinklers) to reduce heat stress and windbreaks to reduce cold stress. In areas of high rainfall, where soils are prone to pugging and pastures become damaged, cows can be excluded from paddocks without damaging their feet and joints as can be the case with concrete; they are often called stand off pads and are discussed later in this chapter. They should be close to the feeding area and milking parlour, otherwise farmers may still prefer to use more accessible paddocks.

With regards size of the loafing pad, cows should each be provided with 9 m^2 and this should be increased to 15 m^2/cow if they live on the feedpad for weeks on end. Smaller areas will require more cleaning and closer attention. Regular cleaning is essential so rainwater can soak through or drain away.

The simplest form of loafing area is a sacrifice paddock, which ideally should be on porous soils on hillsides. Limited shade trees can be a problem as cows tend to congregate under them and this can lead to pugging and mud during wet weather.

The simplest all weather loafing pad consists of a compacted clay base, often with slotted sub-surface drainage pipes 1.5–3 m apart and falling at 1% or more (see Colour plate 3). These slotted pipes are then covered by 20 mm gravel which in turn is covered by 80 mm coarse sand and then 600 mm or more of rice hulls. Geosynthetic material can also be laid over the initial gravel. Pine bark, saw dust and even sand can be used if rice hulls are not readily available. The manure is removed as much as possible from the top layer and more hulls are added when the surface becomes too contaminated. Once every 12–24 months, the hulls should replaced to remove the impervious layer of manure and hulls.

Loafing pads can also be constructed of gravel together with clay or other fine material, such as limestone, which is compacted to set hard. Water will easily drain off these pads by ensuring a 2–4% slope.

Loafing pads are often called loose housing where the cows are free to lie down anywhere on the pad, in contrast to free stalls where cows are allocated specific areas to lie down.

Feed and water troughs

These are discussed in Chapter 8 under feeding management.

Free stall cubicles and sheds

A free stall shed is essentially a feedpad with the addition of specific bedding areas for the stock to lie down. It is generally a covered shed and may include a loafing area for cattle to also be loose-housed where they can stand, ruminate or idle. When well-designed and managed, free stalls provide the ideal system for managing dairy cows intensively off pasture as each animal is provided with a specific place to rest, their management (feeding, cleaning and relaxing) can be optimised and the system can operate efficiently with minimal labour.

They are expensive to construct, however, and can become very unprofitable if stock suffer from poor welfare, animal health and milk quality due to poor feeding and herd management.

Free stalls are individual cow-bedding cubicles where partitions orientate stock for comfort and sanitation, providing each cow with a dry and comfortable place to lie down and rest and ruminate. Free stall sheds should have one stall for each lactating cow. Some farmers provide additional stalls to allow for herd growth and to provide areas for subordinate animals to move away from more aggressive herd mates.

Cubicles or stalls can be arranged in a single row or in more than one row with a central feeding alley or with feeding alleys along the sidewalls. The cubicles can be arranged with cows facing one another (head-to-head) or the other way around (tail-to-tail). With the tail-to-tail arrangement, a central cow alley, 2.2 m wide between the cubicles is needed. If the cubicles are head-to-head, two cow alleys behind each row are necessary. Usually one of these alleys is combined with the feed alley. Free stalls are usually laid out in modules with crossovers providing access to the feeding alley. These can provide multiple routes between cubicles and feeding area and so minimise the adverse effects that dominant stock can have on eating behaviour of submissive stock. Some free stall sheds have self-catching lockable feeding head stalls along the feed line to allow animals to be caught for veterinary attention, insemination or even locked away from the feed.

Stall dimensions should be based on the largest 25% of the herd to allow for increase in cow size through improved feeding and genetics over time. They should also provide for adequate lying down as well as necessary forward and sideways lunging to stand. Typical Friesians require about 240 cm long × 20 cm wide lying space with a further 60 cm forward lunging to allow for normal standing behaviour. Forward lunging space can be shared where two rows of stalls face head to head. Further details on stall design and dimensions are provided by O'Keefe *et al.* (2010).

Stalls that are too long or wide allow the animal to move forward in which case faeces and urine can be deposited within the stall and not in the alleyway. To further prevent cows from soiling the cubicles, shoulder and neck rails are needed to force cows backwards when they stand up. The distance of the adjustable shoulder rail to the back of the cubicle, measured diagonally, should be about 1.8 m, and the height to the cubicle floor may vary between 0.9 m and 1.05 m.

The ideal lying surface is soft, absorbs moisture and does not promote the growth of bacteria. When cows are forced to lie on hard surfaces, they do not lie down for long, are more unsettled and may develop knee and hock lesions and swelling. The stall base is usually made of concrete or compacted fill, which is then covered with bedding, or even a mattress filled with recycled rubber (or crumb, foam or water). Alternatively, mats can be placed underneath the bedding. All base types need loose bedding material on top for further cushioning, moisture absorption and to reduce friction. If the stall base provides good cushioning, less bedding is needed on top.

The simplest bedding is packed earth or sand but it needs care to maintain a flat surface. Sand is quickly pushed around by cows, and should not be used with mechanical or liquid manure-handling systems because it fills up storage tanks and is very abrasive, damaging equipment such as manure pumps.

A concrete foundation with a disposable bedding of chopped straw, sawdust, wood shavings or crushed corn cobs is more common in Europe, as rice hulls are not readily available. Rice hulls would make ideal bedding for free stalls but their high silica content could damage liquid manure handling equipment. Hard surfaces should have a slope of at least 1% so that urine will drain into the cow alleys.

Manure and wet bedding should be removed and replaced with dry bedding material each day. Cleaning should be frequent enough to keep the back of the stall clean because this is where the cow's udder and teats are in contact with the bedding when she lies down. Organic bedding (sawdust, straw, hay, composted manure, rice hulls) should be added every 1–3 days, especially on mattresses and rubber mats, as it is hard to keep bedding on these surfaces.

Dirty cow alleys will result in dirty beds and udders, weakened hoof horn and potential mastitis. Cow and feed alleys should be kept clean by manual scraping, automatic scrapers or flood washing. Although cows can still be in their stalls, it is better to time flood washing when they are away from the feedpad.

Rotating cow brushes are sometimes provided to allow cows to groom and scratch themselves. It may also reduce frustration or stress due to boredom. The free stall environment should be made safe for the stock through ensuring they cannot put their heads through gates and fences or get stuck under stall divisions and barriers. There should be no projections, such as broken boards or rails or rough, sharp edges on the concrete. Rails should be strong enough not to break when cows lean on them. Walking surfaces should be grooved to minimise slips and falls and so encourage normal oestrus activity. Cattle crushes and head gates should be well designed to ensure that stock can be examined without fear of injury.

The free stall facility should be designed to ensure smooth and quiet cow flow. There should be no sudden changes from light to dark, reflections or drains across the cow alleys. Cows will move more smoothly along curved races, up a slight incline and where they have sure footing. Gates could be muffled by attaching rubber strips to prevent excessive noise. Yards must be designed for easy drafting of targeted cows. Stock should only be moved around using 'flappers' (leather strips attached to a cane) rather than using wooden or metal pickets or pipes. Excessive twisting of an animal's tail is unacceptable and electric prods should only be used in emergencies.

To minimise climatic stresses, windbreaks or trees will reduce cold winds while natural inclines and earth can also provide shelter, as can large straw bales, corrugated iron. Walls higher than 2 m will cause eddy currents, directing some wind down onto the cows, to ensure some movement of fresh air. Shed designs for heat and cold stress management are discussed in Chapter 9. Plans for a free stall shed are presented in Figure 4.2. The key considerations in the design of free stall sheds have been summarised by O'Keefe *et al.* (2010).

Other aspects of feedpad design

Air pollution: odours and dust

Feedpad *odours* are produced from anaerobic biological activity occurring during decomposition of manure, spilt feed and other organic matter. Cow numbers, climate, type of feed, duration of feed storage and feedpad management all affect feedpad odour. Buffer distances are a necessary means of reducing the impact of odours. Good feedpad design and management and regular cleaning and maintenance will reduce or virtually eliminate the risk of offensive odours.

Odour is all too frequently one of the major problems encountered in operating feedpads. It is generated from wet manure, urine, stored and spilt feed (particularly by-products), anaerobic holding ponds and by stock. Odour nuisance is very subjective and hard to measure, with no single method available. Odours can be minimised by making sure laneways and feeding areas are well-drained to ensure runoff is directed away from yards and feed areas; maintaining water troughs to prevent wet areas in yards and feeding areas; enclosing feed and water troughs completely to prevent any build up of spilt feed and manure in difficult to clean locations; designing stock and feed sheds with proper orientation and good ventilation; controlling moisture levels, temperatures and aeration of feed storage facilities; locating silage storages away from neighbours; ensuring manure is conveyed in drains from stocked areas to treatment and holding systems; ensuring these drains can be self-cleaned and are kept free of manure; designing solid collection and effluent pondage correctly, and ensuring appropriate desludging frequencies; and maintaining a stock-disposal program to remove dead animals quickly

Other steps to minimise odours during effluent applications are discussed in Chapter 5.

Dust arises from movement of stock on the feedpad and along laneways and the handling of solid wastes (storage, processing and land application). Manure, grain dust and composted material contain fine particles that contribute to dust emissions when these materials are dry. Dust problems can be reduced by designing laneways to be relatively narrow, fenced and well-drained; designing yards to exclude runoff from other areas and regularly flushing them; planting wind breaks around dust sources, such as feed processing areas; designing roads

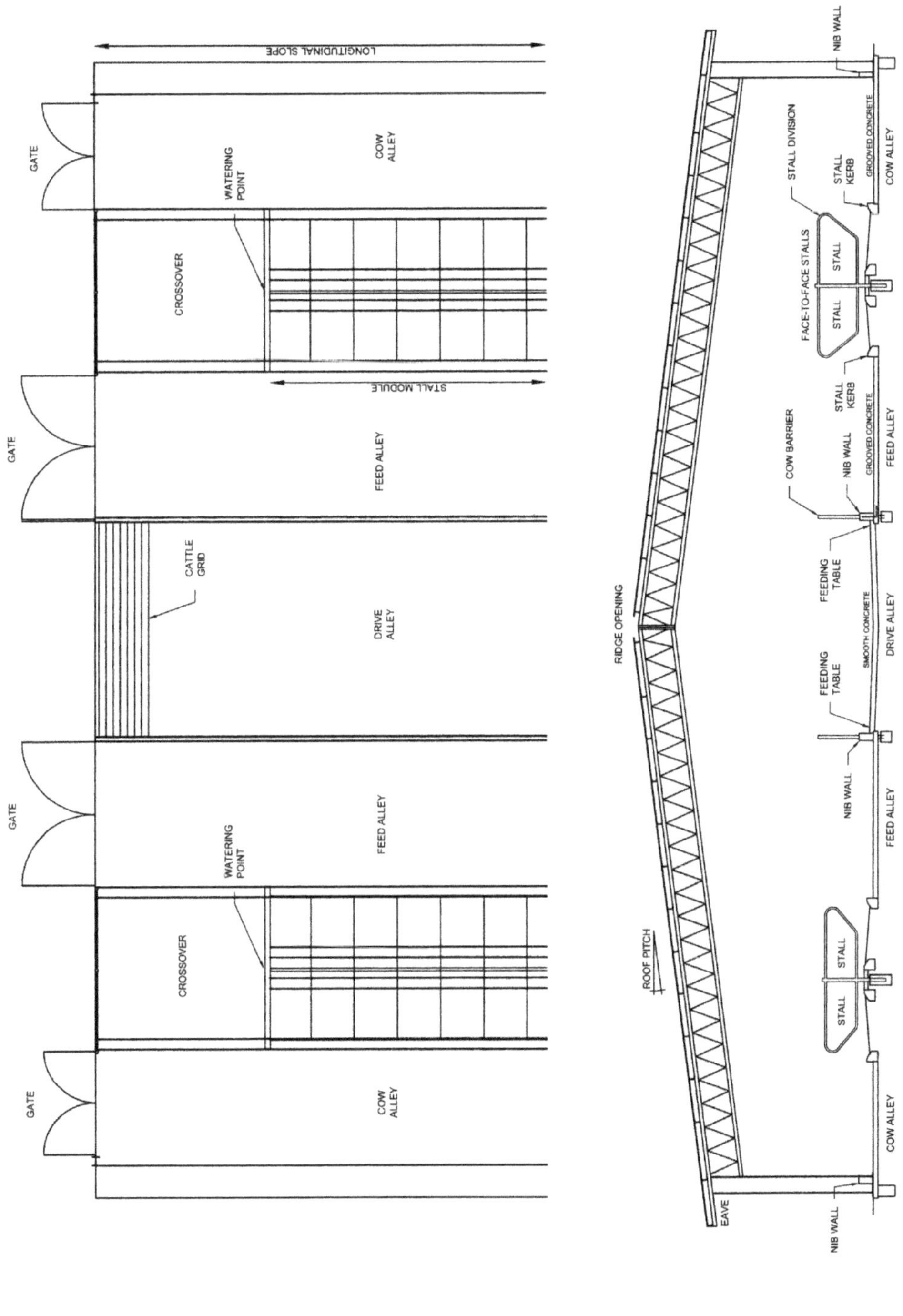

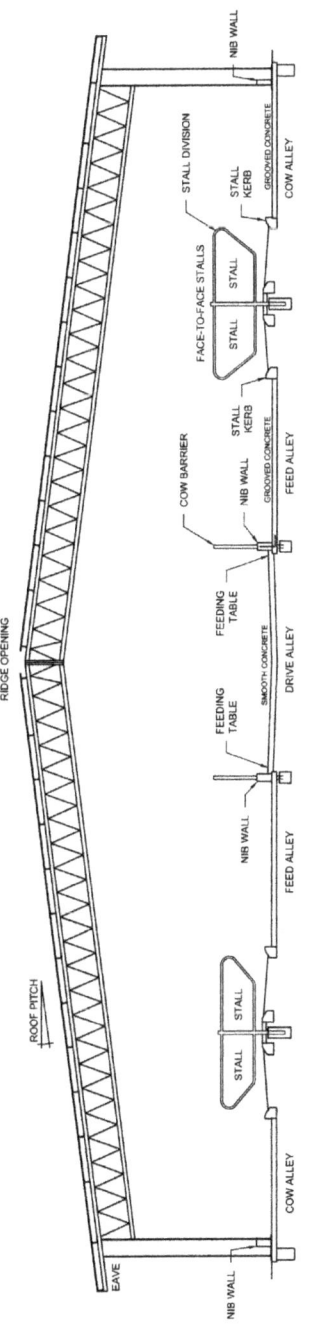

Figure 4.2 Plans and cross-section for a four-row free stall shed.

within the dairy and feedpad areas to minimise dust generation; spraying water on and consolidating dusty surfaces when needed; timing or managing operations involving moving large quantities of solids to minimise dust; and siting solids-processing and storage areas away for neighbours.

Noise

Noise is rarely a problem because of buffer distances to reduce odour problems; however, there are certain points to consider. These include the hours of feedpad operation, which need to be assessed to reduce the impact on nearby neighbours; the delivery of feeds or transport of stock using heavy vehicles, which can cause a problem if they enter or leave the farm between 2200 and 0600 hr. Truck access should be at least 250 m from nearby houses with all vehicles having efficient exhaust mufflers. Advisory routes may be useful for minimising trucks operating in townships. Feed machinery should be properly maintained with all noise-abatement equipment installed while all mechanical equipment should be located and designed to minimise mechanical noise or vibration being identified off-site. Noise generated by machinery should not exceed existing background noise levels between 2000 and 0600 hr, while enclosure and efficient insulation may be required for the feed preparation plant.

Landscaping

This plays an important part in softening the visual aspect of a feedpad. Belts of landscaping could be established around the feedpad to provide visual screens from roads, public areas and nearby residences. Where possible, existing trees should be retained and incorporated into the design. Such vegetation provides wind breaks, lowers the watertable, helping to reduce seepage and water-borne nutrients that have escaped from the site. Trees and other vegetation can be established down slope from the feedpad to assist in filtering any seepage. Such species should be low maintenance and not require further watering following planting and perhaps the first summer. Careful selection is required because of the likelihood of high nutrient levels in the groundwater. Deciduous trees, although harder to establish and maintain than native trees, can aid climate control of the stock by providing summer shade yet not reducing winter sun.

As feedpads are often perceived as feedlots for intensive stock husbandry, community perceptions cannot be ignored. Feedpads can be viewed as point sources of odour, nutrients, noise and pests or insects. In addition, managing stock in areas with high concentrations of manure can be perceived as an animal welfare issue. Increasing incidences of mastitis and other diseases, leg trauma injuries and climatic stress in poorly managed feedpad operations must be addressed quickly to allay such community concerns. With appropriate attention to ensuring an aesthetically acceptable operation with healthy stock and minimal environmental pollution, these perceptions can be overcome. Certainly, if such a facility 'looks good', it is less likely to attract negative attention.

Other aspects of feedpad designs are discussed later (see Chapter 8 for feed and water supplies; Chapter 9 for animal welfare issues, and heat and cold stress management).

The role of a stand off pad

The primary aim of a stand off pad is to protect pastures by reducing pugging during wet weather. This can be provided by a lounging pad constructed in association with a feedpad. The benefits of stand off pads include less pugging of pasture; more grass in spring; higher stocking rates can be supported and cow condition maintained; possible reductions in calf losses (but this depends on stocking density and the condition of the pad surface compared to the paddock); and the ability to maintain grazing rotation lengths. Stand off pads will protect farm drainage work, such as subsoil drainage pipes. They will reduce the need for grazing off-farm (i.e. agistment for dry cows); and will save the farmer time and worry.

Stand off pads can create addition problems such as their high initial capital cost; the time involved in managing the grazing herd; requirements for regular cleaning, effluent disposal and ongoing maintenance; the transfer of soil fertility from pasture to pad, through manure; the unavailability of surface materials, such as sawdust or woodchip bedding; bullying of some stock, such as in calf heifers, due to overcrowding; mastitis if the surface is not kept clean; and reduced production unless extra management skills are provided.

Shelter from the wind may be provided by wind breaks, corrugated iron or trees. Sub-surface pipe drains and moling are common drainage systems with subsurface drainage outlets covered with wire netting to control rodents.

Cows need to lie down for 8 hr/d but would prefer to lie down for longer, up to 11 hr/d. If lying down is restricted, when cows are given access to pastures, they will prefer to lie down rather than graze.

As stand off pads are only used during wet weather, the pad should be topped up with bedding material during the preceding dry period. Dexcel (2005) recommends replacing 1 m^3 of sawdust or 0.5 m^3 of woodchip per animal. Cowpats should be periodically removed during use and following winter, the pad should be scraped down to a firm surface.

Considerations with stand off pads

When the emphasis on planning a stand off pad is cow comfort rather than feeding system, several aspects require considerations. These are:

- **Surface type** – cows' willingness to lie down depends on the surface softness, slipperiness and wetness. On softer surfaces they will lie down sooner than on harder surfaces. If the surface is overly slippery, cows are less likely to lie down due to fear of injuring themselves. Cows are also reluctant to lie down on wet surfaces. The benefits and other factors to consider for six different surfaces for stand off pads are summarised in Table 4.2.
- **Time on the pad** – hard surfaces are best used for only short periods. The longer the time spent on the pad, the greater the stress. Cow stress is characterised by cows hanging their heads and appearing 'tired' during stand off; increased mastitis levels; cows showing excessive stiffness and lameness; a reluctance to lie down on-pad; and cows exhibiting a preference to lie down on pasture instead of graze.
- **Area available per cow** – when cows stand in a yard, say prior to milking, they require 1.2 m²/cow. To allow cows to lie down in a stand off pad, they require at least 3 m²/cow. Recommendations for minimum area per cow on stand off pads of different surfaces are presented in Table 4.3.
- **Frequency of use** – the more frequent the pad is used, the softer the surface should be and the larger the area per cow.

A checklist for planning a feedpad

To aid planning, the following provides a checklist of considerations for feedpad development:

Legislation and guidelines
- What are the functions of the proposed feedpad?
- Where will the feedpad be located and what state or local government authorities have to be notified?
- What current (and anticipated future) regulations must be addressed?
- If there are no regulations, what are the guidelines, if any?
- Is a planning permit required?
- Has a detailed scaled drawing of the feedpad site been prepared?
- Are there any future feedpad developments planned?
- Will native vegetation be removed during construction? If so, has council approval been sought?
- What is the proposed size of the feedpad (expressed in Dairy Cattle Units), now and in the future, and how does this influence government regulations?
- What is the calculated buffer distance and the actual distance to the nearest residence?

Feedpad construction
- Have the most suitable building and construction contractors been selected?

- From where will I source soil for the foundations?
- Are the soil types suitable for the proposed feedpad development?
- What is the proposed surface construction material?
- What slopes are proposed for the feedpad surface?
- What is the depth of the water table beneath the feedpad?
- Will earthworks affect flooding or water discharge from the property?
- What is the average annual rainfall? Is the site likely to be adversely affected by a one-in-100-year flood event?
- What is the actual size of the feedpad (in m^2) and the average and maximum stocking density?
- What are the physical dimensions of associated laneways, gates, feed alleys and other stock and vehicle access areas?
- Arc there any current or future plans to cover part of the feedpad?
- How will the feedpad be cleaned? How will solids be separated from liquid wastes?
- How much solid effluent is expected, say per month? What are the plans for its recycling?
- For flood washing, how much water is required each day and how much of this will be reused?
- For flood washing, what is the flood wash capacity and flow rate?
- If flood washing is considered, is there sufficient nearby land to utilise all the nutrients in the feedpad effluent fully?
- Will all clean rainfall runoff be diverted away from the effluent storage system?
- Will all contaminated rainfall runoff be collected?
- What is the capacity of the effluent storage? Is it sufficient for six months' winter storage of effluent?
- What are the capacities of pumps associated with flood washing and handling effluent?
- Has the orientation of the feedpad been considered to minimise climatic stress?
- Has sufficient feed-storage capacity been planned for?
- Is there sufficient drinking water available for the maximum number of stock during hot periods?
- What length of feed trough will be available per cow?
- Is there adequate fencing planned for sufficient small groups of stock?

Managing the feedpad
- Have production, management and economic issues been addressed fully?
- What equipment is required for feed preparation, feed distribution, feedpad cleaning and effluent management?
- What animal health plans are there for mastitis, lameness and other foot problems?

Table 4.2 Comparing surfaces for stand off pads.

Surface	Benefits	Factors to consider	Management tips
Concrete	Often already available	Risk of injury through slipping	Introduce cows for short periods
	Easy to clean	Cold surface can cause joint/muscle problems	Some farmers use rubber mat overlay to soften surface
	Durable	Cows tend not to lie down	
	Easy effluent management	Often insufficient area available	
	Location often near dairy	Risk of hoof damage, due to stone bruising, hence lameness	
Woodchip	Warm	Availability of woodchips variable	Soiled woodchips can be composted or incorporated into ploughed soils
	Free draining	May need to change frequently	
	Reduced stress and lameness	Capital outlay for pad	
	Cows lie down sooner	Failed drainage can result in odour, mastitis, lameness and less lying	
	Low chance of stock slipping	Need system for effluent collection	
Sand	Cheap if available locally	Cold doesn't encourage lying when wet	Planting shelter trees or building an artificial.
	Soft on hooves	Prolonged periods on sand can wear hooves	Windbreak around the site will improve cow comfort
	Easy to clean	Necessary to skim off top layer of sand twice per season	
		Becomes soupy if drainage insufficient	
		Needs system for effluent collection	

Surface	Benefits	Factors to consider	Management tips
Sacrifice paddock	No capital outlay	Loss of grazing area	Choose a paddock to be renovated, drainage
	Good for short periods of time	Reduced subsequent pasture growth	Improved or cropped
	Not recommended for long term	Turns to muddy (mastitis, lameness, cows won't lie down)	Don't choose paddock nearby roadside or waterway
		Unable to capture effluent	
		Visibility from road can lead to public concern	
Compact gravel	Cost-effective	Turns to mud easily, degrading lane for future use	Construct drains alongside laneway to capture and divert effluent
(laneway)	Low capital outlay	Difficult to capture effluent	
	Cows can be kept near dairy	Stress due to hard surface	
		Cows may not lie down	
Crop	Uses ground already bare	Gets muddy quickly	Use a moveable water trough and bale feeder to prevent cows walking up and down paddock
	Less moving of stock	Can lead to soil compaction	
	No capital outlay	Needs back fence to preserve soil so no mud	During bad weather, offer a second break each day to prevent them walking about
	Retained soil fertility	Unable to capture effluent	
		Feed a longer break so cows can lie on crop rather than on mud	

Table 4.3 Recommended minimum areas on stand off pads of different surfaces, for crossbred cows*.

Surface	Short term	Long term
	<12 hr/d Two consecutive days	>12 hr/d Three consecutive days or more
Woodchip	3.5 m²/cow	5.0 m²/cow
Sand	3.5 m²/cow	5.0 m²/cow
Concrete (no roofing)	3.5 m²/cow	Not recommended
Laneway	3.5 m²/cow	Not recommended
Crop	8.0 m²/cow	8.0 m²/cow
Paddock	8.0 m²/cow	8.0 m²/cow

* Cows permanently on stand off pads, with no grazing available, require a comfortable place to lie and 8 m²/cow plus 1 m²/cow feeding area. They also require a feed face of 0.7 m/cow or 0.3 m/cow if *ad libitum* feeding. (Source: DEXCEL 2005)

- Have all the relevant issues regarding animal welfare been considered?
- Have all the relevant issues regarding staff management and occupational health and safety been considered?
- Have the skills of staff been upgraded to undertake additional tasks associated with working with stock maintained of a feedpad?

A more comprehensive guide to feedpad and free stall design can be found in the *Guidelines for Victorian Dairy Feedpads and Freestalls* (O'Keefe *et al.* 2010).

5

Managing feedpad effluent

This chapter discusses the collection, storing and recycling of feedpad effluent. The main points in this chapter:

- The management of feedpad effluent includes cleaning the pad (by mechanical scraping and/or using water from hydrants or flood washing), storing and recycling of these solids and liquid on farm, while minimising feedpad odours, flies and vermin.
- Every day each cow produces about 40 kg of raw manure which includes 4.2 kg solids. These solid wastes contain nutrients (nitrogen, phosphorus and potassium) which are routinely applied to soils through fertilisers. Recycling effluent will save on fertiliser costs.
- Well-constructed effluent storage ponds require little maintenance. Directing feedpad effluent into an existing pond is not recommended without careful planning and enlargement. Likewise irrigation reuse dams should not be used for feedpad effluent. Separating the solids from liquid prior to storage is a good option.
- There are a series of trigger points indicating that effluent management needs to be addressed and possibly upgraded.

This chapter present an overview of managing the collection, storage and dissemination of effluents from dairy feedpads. Animal manure can be considered either as a liability or an asset. It is a liability so far as contaminating farm resources, encouraging flies and animal health issues, whereas its role in fertilising pastures is an asset.

Much has been written in recent years on this topic but this has mainly from the viewpoints of building contractors and civil engineers. Such details are relevant to the construction of effluent systems and have recently been summarised in an effluent and manure management database (Dairy Australia 2008b). How much of

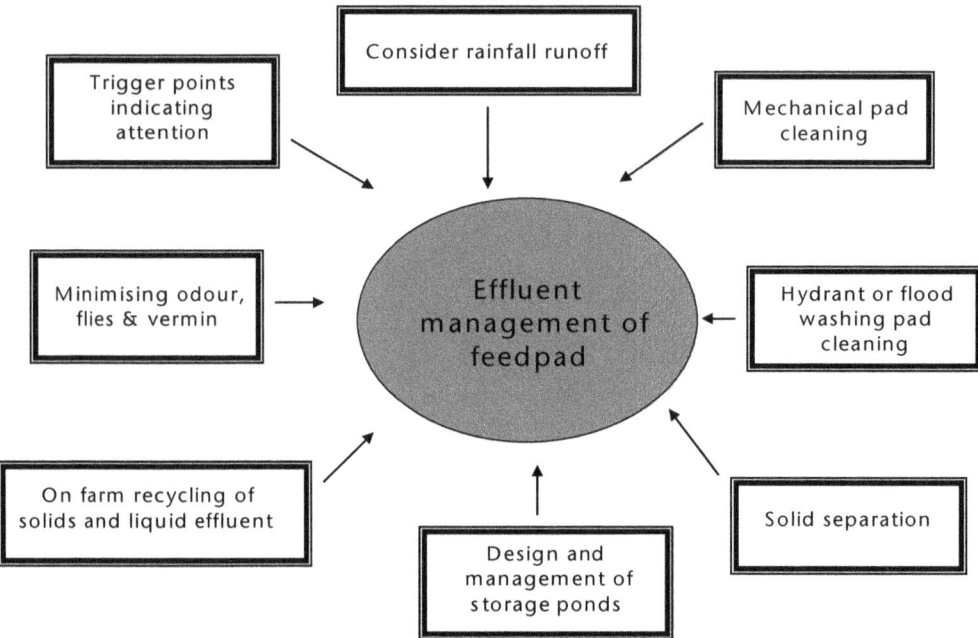

Figure 5.1 The key considerations with collecting, managing and recycling feedpad effluent.

this is of great interest to dairy farmers is questionable. This chapter attempts to repackage the information of most relevance to the owners and managers of feedpads on dairy farms.

An effluent system should be well planned, designed and managed, as it has the potential, if poorly managed, to affect human and stock health and pollute the environment. The key points of effluent design and management are to capture all effluent generated from controlled areas such as feedpads and sheds, and effectively convey it to an appropriate management system; manage the system in a way that the effluent can be reused to utilise the water and nutrients on crops and pasture effectively; manage all effluent in a manner that does not pollute groundwater, surface water or create offensive odours; design and manage the system to suit the site-specific requirements for labour, topography, soils and maintenance; and review and modify the system when farm circumstances change, such as with herd expansion or changes in infrastructure (see Figure 5.1). The feedpad should operate with minimal maintenance input. Soils must be suitable for the construction of effluent storage ponds. Feedpad wastes such as dead animals or silage wrap need to be considered and managed as for the existing best practice for dairy farms.

In summary, from a legislative perspective, all effluent from the dairy, feedpad, stand off areas, underpasses and tracks must be contained and reused, most commonly spread back on pastures and crop. Effluent must not enter surface waters, such as creeks, rivers, billabongs, canals, springs, swamps, natural or artificial channels, lakes and lagoons. Runoff containing effluent must not leave the property boundary. Effluent must not enter ground waters, either directly or through infiltration such as seepage from ponds. Effluent must not contaminate land, such as discharging it onto a small area over a long period and offensive odours must not create an adverse impact beyond property boundaries.

Choosing the most appropriate effluent system requires careful planning. The system must suit the farm's specific characteristics and management strategies. The fundamental principles of any effluent system should include minimising the effluent, as the more generated, the larger the storage capacity and the longer it will take to empty; separating solids and debris, to reduce potential pump damage, conveyancing difficulties and pond desludging; conveyance to the pond, utilising gravity or pumps; storage, or containment of effluent over the wetter months to eliminate any nutrient runoff from the farm or likelihood of pugging pastures; the application to pastures, to distribute effluent and its nutrients back to pastures or forage crops using pumping or gravity irrigation. The more effectively effluent is distributed around the farm, the less the risk of groundwater pollution.

Environmental concerns are increasing and penalties for pollution through ineffective effluent systems will only rise in the future as dairy farming becomes more intensive. As the collection and disposal of dairy feedpad effluent requires specialist skills, it is highly desirable that professional assistance be sought, if only to ensure that the existing (pre-feedpad) system can cope.

Figure 5.2 Feedpad effluent must be carefully controlled.

What is feedpad effluent?

Effluent and feedpad wastes include excreted manure and urine; wash-down water; rainfall that is not diverted off the feedpad; and waste feed and waste bedding.

The total solid (TS) content of as excreted manure may range from less than 5% to more than 20%. O'Keefe *et al.* (2010) differentiate manure as follows:

- Liquid manure: <5% TS.
- Slurry manure: 5–10% TS.
- Semi-solid manure: 10–20% TS.
- Solid manure: >20% TS.

Unless solids are frequently removed from the feedpad surface, they could all end up in the effluent holding pond. The pad is typically scraped with a front-end loader or blade with the solids reused directly or stockpiled within a controlled drainage area for later use or composting. After adding yard and plant wash, the concentration in the waste water is diluted even further, usually to 0.5 to 2% TS.

Feedpad effluent usually has a higher solid, hence a higher nutrient content than dairy shed effluent as feed wastage is combined with wash down waste and cows generally spend a lot longer on a feedpad than in a milking shed. The solids and nutrient content of feedpad effluent depend on time spent by stock on the feedpad; volume of water used to wash down the pad; amount of dry manure and feed waste removed before wash down; feed type and quantity fed; volume of stormwater that is not diverted off the pad; and pre-treatment of effluent prior to storage.

Typical dry matter (DM) contents range from:

- 20% for solids, when separated from feedpad effluent.
- 15% for dry scraped manure.
- 4% for slurry.
- 0.8% for dairy shed effluent.
- 0.3% for liquids in feedpad effluent following solid separation.

The amount of solid manure generated for a dairy herd can be calculated on default values for a 500 kg cow fed on harvested feed. This is 40 kg/d of raw manure or 4.2 kg/d of solids. These amounts are directly proportional to live weight and time that the stock are on the feedpad. Table 5.1 presents estimates of typical volumes of manure loads (adjusted to 8% DM).

Pad-cleaning systems to remove effluent and manure

The type of pad dictates the method of cleaning hence the ease and cost of cleaning. An earth or gravel surface is only suitable for mechanical scraping,

Table 5.1 Manure loads generated by 500 kg cows on feedpads (m²/d).

Herd size	Time spent on feedpad (hr/d)			
	4	8	12	20
150	1.0	2.0	3.0	5.0
200	1.3	2.7	4.0	6.7
250	1.6	3.3	5.0	8.3
300	2.0	4.0	6.0	10.0
400	2.7	5.3	8.0	13.3
500	3.3	6.6	10.0	16.7

(Source: O'Keefe *et al*. 2010)

whereas a concrete pad can utilise both mechanical scraping and flood washing. Where the waste is relatively dry, mechanical cleaning may be sufficient.

Mechanical scraping

Mechanical scraping is the cheapest method and is the only option for non-concrete surfaces. The manure remains dry, hence is easier to handle than if wet. Tractors require cleaning and more maintenance and the job often does not get done frequently enough. As steel blades wear out rapidly on concrete surfaces, it is better to use a rubber-tipped blade.

Figure 5.3 Flood washing a free stall shed.

Manure from the pad surface can be scraped on to a compacted area nearby. Ideally this area should be within the controlled drainage area since seepage from the stockpiled manure needs to be contained and managed.

Scraping works well in the drier months of the year when the deposited manure has an opportunity to air dry. During the wetter months, the feedpad surface can become hard to scrape. The deposited manure can become very sloppy, making it difficult to scrape and stack physically. Adequate feed alley slope (2–4%) will assist in draining earthen and concrete pads during the wetter months.

Although scraping is laborious, it is relatively cheap in capital outlay because most farms will already have a tractor.

Using hoses

High-volume low-pressure hoses can be used to deliver the water. Hosing is time-consuming and difficult with long hoses, and must be a regular routine as the manure is difficult to clean once it dries out.

Flood washing

Flood washing, that is a sudden release of large volumes of water from a holding tank through a series of butterfly valves and pipes, directs the water down the feedpad surface (see Figure 5.3). This water can be recycled for further use. The important variables are volume and flow rates across the width of the feedpad surface. Pre-wetting the surface with sprinklers or soaker pipes assists in manure removal thus allowing the effective use of smaller pipes and flow rates.

When planning flood-washing systems, consider whether the pad should be scraped first; the method of water discharge; the volume of water required, which depend on the amount of solid waste to be removed; the velocity of water during flushing; the control of the runoff, such as size of pipes and other structures; the use of sediment and solid traps; the reuse of water; and the concentration of salts and nutrients which accumulate over time with reuse.

Flood-washing systems can consist of tanks at the highest end of the alley mounted at ground level or on stands to generate sufficient head. The tanks can apply the wash volume directly onto the pad surface through large-diameter piping or the piping can be buried using risers to bring it to the surface. Alternatively, large-volume irrigation pumps can be used to direct freshwater or recycled effluent directly onto the alley or through risers, thus negating the need for a tank.

When using flush pumps to recycle water directly from an effluent storage pond onto the feedpad, pond size and sludge accumulation can be reduced by partial solids removal either with a trafficable solids trap or with mechanised solids-separation equipment (see Figure 5.4). Sedimentation basins also can be constructed before the effluent pond to reduce accumulation of sludge in the long-term storage area.

The washing frequency is determined by the length of time stock spend on the feedpad, hence the quantity of manure produced. For a free stall shed housing cattle for 24 hr/day, the alleys should be washed at least three times daily. Feedpads providing only a portion of the cows' diet are usually washed twice each day once the stock are out grazing or in the loafing area.

O'Keefe *et al.* (2010) provide detailed calculations of typical flood wash volumes required for various feedpad scenarios. A 500 cow feedpad holding cows for 6 hr/d requires 69 000 L/d whereas a free stall shed holding 1000 cows for 24 h/d requires 480 000 L/d of wash-down water.

Table 5.2 provides a comparison of different methods of cleaning.

Table 5.2 Cleaning feedpad surfaces.

Method	Advantages	Disadvantages	Management tips
Scraping	Low capital cost	High labour cost	A 3 m wide scraper blade can be quick and effective
	Less effluent to handle	More concentrated effluent to handle	Clean three to four times each week
			A sprinkler system to pre-wet concrete prior to use aids cleaning
Hosing	Low capital cost	Higher labour cost	Have several hydrants around feedpad
	Large volume of more dilute effluent to handle	Splashing of feed area	
	Uses similar amounts of water to flood washing	Produces large volume of effluent	
Flood washing	Lower labour cost	Higher capital cost	Wide channels of manure accumulations may lead to meandering of flush water. To prevent, divide wide channel into multiple narrow channels
	No splashing of fed area	High discharge rates (12–15 m^3/min) with large diameter pipes (300 mm) needed	
	Uses similar amounts of water to hosing		

Figure 5.4 Solids can be separated out prior to ponding the liquid effluent.

Handling and storing feedpad effluent

Some feedpad systems generate little liquid waste, while others generate a lot. Open drains with a minimum slope of 0.5% are ideal for conveying feedpad effluent. Large-diameter pipes are also suitable, with the larger the diameter, the lower the slope. Grated pits, pipes with less than 100 mm diameter should be avoided. Where possible, gravity should be used to move effluent around the farm because an effluent system reliant on a pump is prone to problems. All systems must have three day sump storage capacity to cater for pump failures and wet conditions.

Taking rainfall runoff into account

All of the liquid effluent should be accounted for when designing the feedpad drainage and storage systems. Rainfall runoff and flood washings are the two major sources that will require storage over months when they cannot be reused for irrigation. Rainfall runoff calculations for the feedpad surface and roof if not diverted and associated work area such as laneways, feed storage and loafing areas, should be based on a one-in-20 year 24-hour storm event using official rainfall and runoff statistics.

The storage volume required is calculated from the following variables:

• The total area of the feed pad and associated facilities

- The one-in-20 year rainfall event. For dairy regions in Victoria, this varies from 65–100 mm/d.
- The runoff coefficient for the feedpad surface. This differs between dirt and concrete and also with annual rainfall.
- The runoff coefficient for any roof over the feedpad or associated work area
- The area of liquid storage.
- A safety factor of 125%.
- If the feedpad is flood washed, the volume required can be easily calculated.

The effluent storage volume needs to be determined for the entire period that wastes have to be stored, such as over winter for an irrigation reuse system. No additional volume is allowed for the manure collected as it can be assumed that this will be offset by evaporation from storage and during the flood washing. If this is already available on farm, then no additional storages are required. Any off farm discharge via regular tanker transport should not be contemplated. An example calculation of required effluent storage capacity is provided by SIRIC (2002).

Solid separation prior to pondage

It is debatable whether it is best to develop a solid separation component on feedpads to remove solids and debris to pondage or simply to direct everything into the pond and deal with it there by desludging. The general rule of thumb suggests if the system is reliant on pumping or if ponds are used in recycling situation for flood washing, it is preferable to remove solids before they reach the ponds. The most common complaint about solids separation is the high labour input and having the right equipment to handle solids.

There are numerous methods to separate solids. It can rely on gravity, such as trafficable solid traps, sedimentation basins and ponds, or use mechanical systems with screening or pressing. It is a matter of selecting the most appropriate to suit the feedpad scale and management of the farm. Designs for solid separation and debris removal can range from simple sump/pump structures, trafficable solid traps or the more advanced pump and screen separators and screw presses, usually more applicable on larger dairy and feedpad operations.

In summary, solid-separation systems offer the following advantages in that they minimise the need for agitation in sumps and reduce the likelihood of blockages in pipes and pumps. They reduce the rate of sludge accumulation in ponds, thus allowing for smaller ponds or requiring less frequent emptying. They allow for conventional irrigation systems to distribute the effluent, and they concentrate the organic matter for transport, composting or direct application to pasture.

Separated solids, having a higher TS content (<20%) will have to be managed separately, thus requiring additional labour, fuel and repairs and maintenance for both their collection and disposal.

Using the existing effluent system

A common practice with dairy feedpad development is to use the existing effluent system on the farm, which services the dairy. For the cost of an additional pump, there is an effluent system ready to go. Unfortunately, in most cases, it is not that simple.

It is very easy to complicate a system that is working sufficiently in the past by increasing the demands to a level for which the system was not designed. Dairy shed effluent systems are generally designed to cater for 10–15% of the cow's daily manure loading as well as the total water generated at the dairy. The time cows spend on feedpad varies considerably ranging from 2–12 hr or more per day, depending on farm management and the time of year. It is very easy to overload effluent systems and reduce their anaerobic functioning by incorporating an intensive feedpad development, without first checking to see if it is a viable option. A few simple questions need to be answered:

1. Is the current effluent pond servicing the dairy shed, already crusted and heavily sludged? If yes, the ponds are already undersized.
2. Does the sump or solid trap at the dairy have enough capacity to cope with extra water use?
3. Is the existing pump sufficient to convey additional effluent to the storage ponds?
4. Have the ponds been sited to maximise access to pasture to ensure an even distribution of nutrients? The more manure deposited into the ponds, the greater the area required for nutrient distribution.
5. Will there be more pipe blockages?

For the effluent system to be adequate, it must be capable of holding all the effluent and water used at the dairy and on the feedpad, taking into account any rainfall runoff that can enter the effluent stream. For example, the amount of effluent produced over 24 hr on a feedpad can be equivalent to a week's output from the milking shed. It must be remembered that it is not the herd size that determines pond sizing; it is the amount of water used in washing the facility.

Recycling effluent water for feedpad cleaning

A two or multiple pond system is usually recommended for effluent to be recycled back to a concrete feedpad for washing alleyways, whether it is through a series of hydrants, wash down hose or flood wash system. Dual ponds provide better quality water, therefore reducing the likelihood of pump breakdown and odour problems.

The ability of microbes to break down the manure can also be affected by the accumulation of salt within the pond. This is usually the result of poor quality wash down water or irregular emptying of the pond. Even multiple ponds systems designed for effluent recycling need emptying periodically.

Irrigation reuse dams

Irrigation reuse dams are not designed for effluent storage and therefore effluent should not be conveyed directly from the shed to the reuse storage. Nutrient build-up in reuse dams promotes the growth of blue-green algae, and this is more likely to occur if dairy effluent is allowed to enter the reuse system. There is also a higher risk of effluent leaving the farm via external outlets during a heavy rain event.

Effluent storages need to be closed systems with no direct access to irrigation reuse sumps or dams. Under special circumstances the reuse dams may be used as method for effluent conveyancing to pastures; however, this tends to place a high risk on farm management to maintain nutrients on farm.

Drainage

Runoff from feedpads and intensively stocked areas contains relatively high concentrations of many contaminants. Therefore, irrespective of size or stage of development of the feedpad, provision must be made for drainage and pollutant control.

Rainfall is a significant factor, which must be considered in the design and management of feedpads, especially exposed surfaces with heavy stock access. The protection of surface and ground water by preventing nutrient runoff and leachates must be assured. Good drainage is essential for the maintenance of disease-free stock and high quality milk. It also provides a management system to control nutrients and redirect them back to pastures.

Pond maintenance

A well-constructed and designed pond will require little maintenance, especially in a single pond system, which is operating anaerobically. A correctly functioning anaerobic pond will not crust, and rising gases indicate that rain is falling on the pond.

For storage ponds to work effectively they must be sufficiently emptied at the start of the wetter months, so that all effluent produced over that period can be retained in the pond. Ponds should be constructed to ensure there is always 500 mm depth of effluent or sludge on the bottom. This will protect the base of the pond from cracking and also maintain the anaerobic process. Ponds that form crusting indicate that they are undersized for the daily loading generated at the dairy and/or feedpad. Ponds that crust continually will require more frequent maintenance. The preferred option to address this problem is to enlarge the pond, or improve solid separation prior to storage. Indicators that a pond requires maintenance include an overflowing pond; heavy crusting on the surface; excessive weed growth on or around the pond; reduced storage capacity due to sludge build-up; evidence of seepage; bank and batter erosion; problems with pump and conveyance pipes; the existence of strong odours; bubbling has stopped in the anaerobic pond (the first of a two-pond system); or there is severe discolouring in any nearby waterways.

Managing effluent odour

It is important to manage effluent systems to reduce the risk of offensive odours impacting offsite. Consider wind direction when agitating, desludging or applying effluent sludge to pastures and crops. Establish and maintain suitable distances from neighbours when applying or storing effluent. Avoid dumping large quantities of milk in effluent systems as it reduces pond performance and creates odours. Milk should be disposed of in accordance with a contingency plan. Reduce boggy and damp areas and laneways with appropriate drainage and management practice. Remove spoilt and wasted feeds from intensive feeing area and incorporate into manure stockpiles. If sludge removal releases strong odours, manage the process to minimise them and advise neighbours of when it is planned. When disposing of effluent on farmland, consider application rates hence residues remaining on soil surface, the prevailing and forecast weather. Avoid weekend applications if possible. Create piles of solid waste on low permeability soils on a slightly sloping surface. Maintain a dry surface and prevent water logging of these solid wastes. Redistribute the solid wastes around the farm as soon as possible after collection and during the middle of the day, when odours can be quickly dispersed. Consider other methods of effluent application such as irrigation (spray, surface, drip) tanker spreading and deep injection. These may not be viable alternatives.

Minimising flies and vermin

There are a range of actions that can be taken to minimise flies and vermin, such as using appropriate baiting, trapping and spraying practices to control outbreaks; reducing excessive weed and grass growth around ponds and feed storage areas; maintaining extra cleaning schedules to remove manure deposits around yards and access ramps; ensuring the most appropriate herd health practices are used, such as insecticide repellents; avoid disposing of carcasses in and around the pond area; and record and erect signs at baiting and trapping stations.

Recycling feedpad effluent

Feedpad effluent provides maximum benefit when used to grow crops. Even though the site selected should have sufficient area to utilise all water applied, it is likely that the nutrient content of the water will dictate the area of land required. Feedpad effluent systems should not be directly linked to existing irrigation reuse systems, although provision could be made for period transfers. Obviously the land should be on-farm, but if off-farm, attention must be paid to the transport of the wastes and transfer of diseases between farms.

Machinery is available for spreading solid effluent onto land, whereas liquid effluent, when not recycled for flood washing, must be applied by an approved irrigation method and not discharged continuously from one point. It can be mixed with flood irrigation water to be applied. In addition, the Biological Oxygen Demand (BOD) and the presence of pathogens in the waste water, should be considered. The BOD is the amount of oxygen required to stabilise decomposition of organic carbohydrates under aerobic conditions.

Effluent as fertiliser

Incorporating effluent reuse on farm will change the overall nutrient management, hence the fertiliser program for the farm. Effluent contains nitrogen (N), phosphorus (P) and potassium (K) which will reduce fertiliser requirements, as well as sodium and potassium salts. Nutrient budgeting should then be used to balance the supply and demand of the nutrients in the farm's overall cropping program. Crops should remove all the nutrients applied to prevent a build up of nutrients in the soil and the possible contamination of the groundwater. Alternatively, a minimum default value of 1 ha/5 cows can be applied if the stock are always located on the feedpad, with adjustments made for actual time on the pad. Annual soil testing and monitoring of groundwater should indicate any undesirable build-up of soil nutrients.

Figure 5.5 Large water tank for flushing effluent off a feedpad.

The default values for farm nutrients generated for a 500 kg dairy cow, producing 40 kg/d of raw manure or 4.2 kg/d of solids (SIRIC 2002), are:

- Nitrogen (N): 0.225 kg/d
- Phosphorus (P): 0.047 kg/d
- Potassium (K): 0.145 kg /d

Therefore a 500-cow herd would annually generate 41.1 t N, 8.6 t P and 26.5 t K. The economic benefits of effluent replacing fertilisers are discussed in Chapter 11.

Animal health issues

There are several animal health issues to consider in preventing and minimising potential infection and spread of disease, such as excluding stock from around effluent ponds and drains; preventing access to any pastures with effluent applied for young dairy stock less than 12 months of age, to minimise the likelihood of Johnes Disease; withholding periods of at least three weeks after an application of effluent, are recommended before grazing, to avoid nitrate and palatability problems. In addition, cows should not be allowed to calve down on areas with significant applications of effluent. Farmers should monitor effluent application areas for soil fertility targets to minimise potential herd health problems, such as high potassium levels. They should provide stock with clean water from bores or tanks and avoid water that may have been contaminated by effluent; ensure drainage networks from cattle grazed paddocks do not impact upon calf rearing areas; establish a grazing management plan to ensure effluent applications do not compromise paddock rotations. Effluent-handling equipment should be cleaned and disinfected between farms to reduce the spread of disease.

Triggers to address effluent management

It is one thing to know the principles of efficient effluent management, but there is nothing better than practical experiences to anticipate problems before they occur. Below are some of the trigger points that can indicate when effluent management needs to be assessed closely and possibly upgraded:

- Once water is used for cleaning feedpad surfaces.
- When the feedpad is significantly enlarged, such as the building of a loafing area.
- When it is located close to a 'sensitive receptor' such as waterway or neighbour's house.
- When the herd occupies the feedpad for longer periods, for example if it becomes used regularly as a winter stand off pad.

- When herd health becomes compromised, such as increases in the incidence of mastitis.
- When the neighbour starts complaining; for example, if odour becomes an issue.
- When soil testing indicates 'hotspots' or areas with very high soil nutrient levels, associated with effluent application.
- When sand is being reclaimed for bedding, say for free stall sheds, or it becomes a problem during solid separation.

6

Key principles of dairy nutrition

This chapter explains the key principles of dairy nutrition to help improve understanding of the practices of feeding management.

The main points in this chapter:

- Feeds contain the nutrients for animal survival and production, the most important ones being water, energy, protein and fibre.
- Nutritionists measure feed energy in terms of megajoules of metabolisable energy per kilogram of dry matter (MJ/kg DM), protein in terms of % crude protein and fibre in terms of % neutral detergent fibre.
- It is possible to calculate the nutrient requirements of milking cows to produce a target level of milk.
- Understanding the lactation cycle from one calving to the next is important when planning and managing the herd's annual feeding cycle.
- Full lactation milk yields depend on peak milk yield and the rate of decline of milk yield from peak (or lactation persistency). A realistic target persistency over the full lactation would be 7–8% decline per month from peak.
- There are many factors influencing the amount of additional milk that cows produce in response to the feeding of extra supplements. One important influence is the substitution of pasture for supplement in that cows generally eat less pasture as they are offered more supplements.
- Some of the milk response is immediate, while some is delayed until the following lactation.
- Decisions to supplement grazing cows should be based on the marginal milk response; that is, the supplement costs and milk returns arising from the next level of supplement fed.
- There is a large discrepancy between the theoretical and actual milk responses derived from feeding cereal grains to grazing cows.

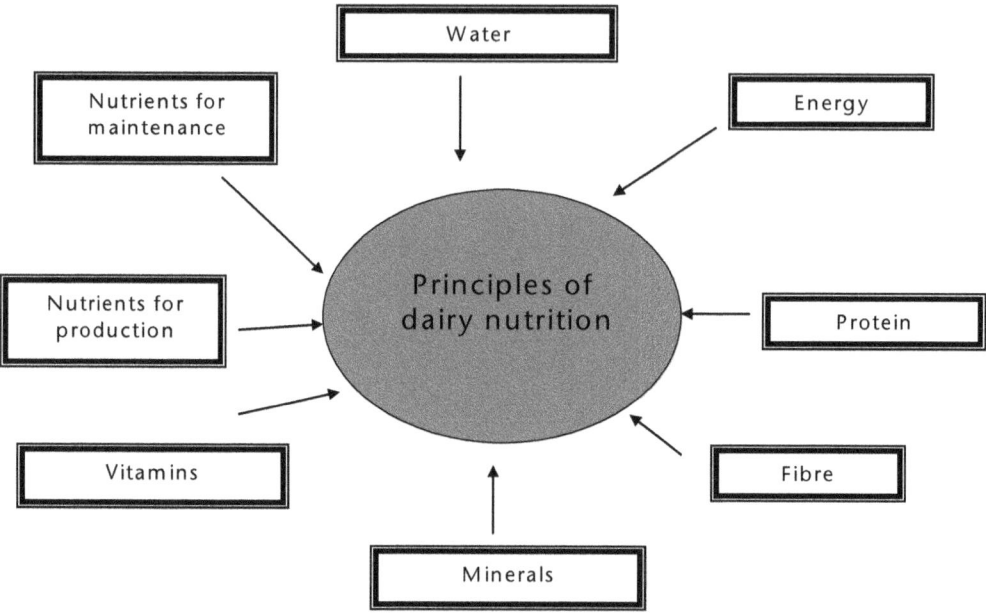

Figure 6.1 The key feed nutrients for milking cows.

One of the major benefits of feedpads is the ability to provide a range of supplements for grazing stock with reduced wastage that would occur if fed out in the paddocks. It is then important for farmers to balance the requirements of grazing milking cows with feedpad supplements. Depending on pasture supplies, feedpads can be used to supply anything between 5% and 100% of the stock's daily requirements.

Feeding management includes both the supply of feeds and their presentation to the milking cow. This chapter discusses the processes involved in calculating how much feed nutrients are required to achieve milk production targets while Chapters 7 and 8 discuss some of the logistics of supplying these feed nutrients.

The digestive system of the dairy cow

Dairy cows are herbivores and have digestive systems well adapted to forage-based diets. Belonging to a group of mammals called ruminants, they have a multi-purpose digestive system. As well as the stomach that breaks feeds down into their basic nutrients, they have a rumen which is like a large fermentation vat where micro-organisms pre-digest these feeds. The rumen allows dairy cows to make use of feeds that would otherwise be wasted if consumed, through the microorganisms (or microbes) living in the rumen. Therefore the approach to feeding dairy cows is 'look after these microbes and they will look after the cow'.

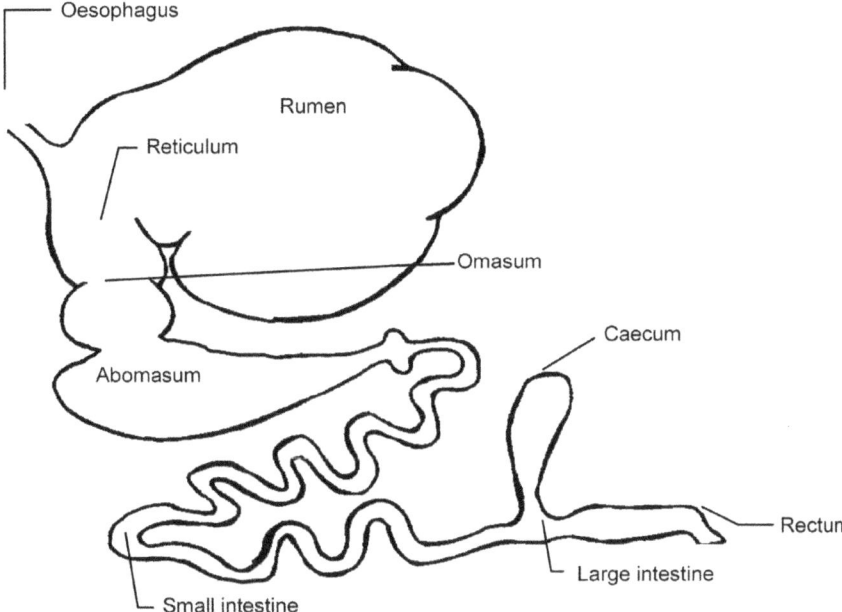

Figure 6.2 Digestive system of the dairy cow.

Three steps are involved in cows obtaining nutrients from their diet:

- Ingestion: taking food into the body.
- Digestion: food is mechanically and chemically broken down.
- Absorption: nutrients pass from the digestive system into the cow's bloodstream.

The digestive system of dairy cows is well adapted to a forage based diet. As ruminants, cows have one true stomach (the abomasum) and three other compartments (the rumen, the reticulum and the omasum) which each have specific roles in the breakdown of the feed consumed. These are shown in Figure 6.2.

The rumen and reticulum

Once food has been ingested, it is briefly chewed and mixed with saliva, swallowed, then it moves down the oesophagus into the rumen. The rumen is adapted for the digestion of fibre. It is the largest compartment of the adult ruminant stomach. Its internal surface is covered with tiny projections called papillae which increase the surface area of the rumen and allow better absorption of digested nutrients.

The reticulum is separated from the rumen by a ridge of tissue. Its lining has a raised honeycomb-like pattern, also covered with small papillae. The rumen and reticulum together have a capacity of 50–120 L of food and fluid. The temperature inside the rumen remains stable at around 39°C (38–42°C) which is suitable for the growth of a range of microbes.

The microbes break down feed through the process of fermentation. Under normal conditions, the pH of the contents of the rumen and reticulum is maintained in the range of 6–7. It may be lower in grain-fed cows. The stable pH range is maintained by continual removal, via the rumen wall, of acidic end products of microbial fermentation, and by the addition of bicarbonate from the saliva.

Saliva

Saliva has several roles: it makes chewing and swallowing easier, but primarily it contains sodium (Na) and potassium (K) salts that act as buffering agents against rumen acidity.

A cow can produce 150 L or more of saliva daily. The volume of saliva secreted depends on the time spent eating and ruminating.

Chewing and rumination

Before food reaches the rumen its breakdown has already begun by the mechanical action of chewing. Enzymes produced by the microbes in the rumen initiate chemical breakdown. The walls of the rumen and reticulum move continuously, churning and mixing the ingested feed with the rumen fluid and microbes. The contractions of the rumen and reticulum help the flow of finer food particles into the next chamber, the omasum.

Rumination, or chewing the cud, is the process whereby newly eaten feed is returned to the mouth for further chewing. This extra chewing breaks the feed down into smaller pieces, thereby increasing its surface area making it more accessible to the rumen chemicals. As a result, the rate of microbial digestion in the rumen is increased.

The time spent ruminating (chewing the cud) depends on the fibre content of the feed. The more fibre in the feed, the longer the ruminating time, therefore the less feed that can be eaten overall, and the less milk produced.

Some nutrients are absorbed across the rumen wall. Absorption involves the movement of individual feed components through the wall of the digestive tract into the bloodstream where they are transported to the liver.

There is a constant flow of digesta through the digestive tract. Because food larger than 1 mm cannot leave the rumen until its length is reduced, the rumen is the major regulator of feed intake.

Passage of food through the rumen

The passing of material through the rumen affects the extent of digestion. General rate of passage depends on density, particle size, ease of digestion and level of feeding. Some foods pass through the digestive system fairly quickly, but very indigestible food will take a longer period.

Microbes of the rumen and reticulum

The microbes in the rumen include bacteria, protozoa and fungi. These microbes feed on forages ingested by the cow, and by fermentation, produce end-products that are utilised by the cow as well as by the microbes themselves for their own reproduction and cell growth.

Bacteria and protozoa are the most important microbes. Billions of bacteria and protozoa are found in the rumen. They digest about 70–80% of the digestible DM in the rumen. Different species of bacteria and protozoa perform different functions. Some digest starch and sugar while others digest cellulose.

The numbers and proportions of each type of microbe depend on the animal's diet. Maintaining a healthy mixture of different microbes is essential for keeping the rumen functioning efficiently.

The major end-products of microbial fermentation are volatile fatty acids, the products of fermentation and the cow's main energy source and ammonia, which is used to manufacture microbial protein. Bacteria are 60% protein, making them the major source of protein for the cow as they leave the rumen and are digested in the abomasum and small intestine and gases, sources of wasted energy as they are belched out regularly.

Dietary upsets, such as feeding too much grain too quickly, can cause a rapid change in the microbial population. This changes fermentation patterns and interferes with fibre digestion. Varying the level of grain fed should therefore be done gradually so that the populations of rumen microbes can adjust accordingly.

The speed of digestion of feeds depends on the quality and composition of the feed. It is affected by the number and type of microbes, the pH in the rumen, the nutrients limiting the growth of the microbes and the removal of microbes from the rumen. Energy and protein are the major nutrients, which influence microbial growth and therefore rumen fermentation.

The microbial population needs energy and protein for growth and multiplication. If either of these nutrients is in short supply, microbial growth is retarded, and so is the rate of digestion (the digestibility) of feed.

The omasum and abomasum

The omasum lies between the reticulum and abomasum. The material entering the omasum is made up of 90–95% water. The primary function of the omasum is to remove some of this water and to further grind and break down feed. Large plate-like folds known as laminae extend from the walls of the omasum. These folds are attached in the same way as pages are bound to the spine of a book.

The laminae are covered in papillae which direct the flow of food particles towards the next chamber, the abomasum.

The abomasum connects the omasum to the small intestine. Acid digestion, rather than microbial fermentation, takes place in the abomasum, much the same as in the human stomach.

The lining of the abomasum is folded into ridges, which produce gastric juices containing hydrochloric acid and enzymes (pepsins). The pH of these gastric juices varies from 1 to 1.3 making the abomasum very acid, with a pH of about 2.

The acidity in the abomasum kills the rumen microbes. The pepsins carry out the initial digestion of microbial and dietary protein in the abomasum.

The small and large intestines

From the abomasum the digested food moves to the small intestine. There, enzymes continue the digestion of feeds and microbes. Most nutrient absorption occurs in the small intestine.

The large intestine, mainly the caecum and colon, is the site of secondary fermentation, particularly of fibre. About 10–15% of the energy used by the cow is absorbed from the large intestine. Absorption of water, minerals and ammonia also occurs here. The components of feed not digested in the large intestine pass through to the rectum then expelled as faeces.

Nutrients supplied by feeds

When cows consume their smorgasbord of feeds, the nutrients they extract from them are water, energy, protein, fibre, minerals and vitamins (see Figure 6.1). The four nutrients that have the major influence on cow performance are water, energy, protein and fibre.

1. Water. The body of a dairy cow is composed of 70–75% water. Milk is about 87% water. Water is not a feed as such because it does not provide specific feed nutrients. It is essential, however, in body processes and to regulate body temperature. Water is involved in digestion, nutrient transfer, metabolism and waste removal. Water has structural and functional roles in all cells and all body fluids. An abundant, continuous, and clean source of drinking water is vital for dairy cows.

2. Energy. Dairy cows use energy to function (walk, graze, breathe, grow and put on body condition, lactate, and maintain a pregnancy). Energy is the key requirement of dairy cows for milk production. It determines milk yield and milk composition.

3. Protein is the material that builds and repairs the body's enzymes, hormones, and is a constituent of all tissues (muscle, skin, organs, foetus). Protein is needed

Figure 6.3 Comparing free stalls and loose housing at DPIV Kyabram.

for the body's basic metabolic processes, growth and pregnancy. Protein is also vital for milk production.

Proteins are made up of nitrogen which are bound into various amino acid molecules. Amino acids are the building blocks for the production of protein for milk, tissue growth and the development of the foetus during pregnancy.

4. Fibre. For efficient digestion, the rumen contents must be course with an open structure and this is best met by the fibre in the diet. Fibre contains most of the indigestible part of the diet. Cows require a certain amount of fibre for rumen function. It ensures that the cow chews its cud (ruminates) enough and therefore salivates. Saliva buffers the rumen against sudden changes in acidity.

5. Minerals are inorganic elements. They are needed for teeth and bone formation, enzyme, nerve, cartilage and muscle function or formation, milk production, blood coagulation and efficient utilisation of energy and protein.

Some minerals (macro-minerals such as calcium and phosphorus) are required in large amounts while others (trace minerals such as copper and zinc) are only required in minute amounts.

6. Vitamins are organic compounds that all animals require in very small amounts. At least 15 vitamins are essential for animals. Vitamins are needed for many metabolic processes in the body; e.g. for production of enzymes, bone formation, milk production, reproduction and disease resistance.

Describing feed energy

There are various ways to describe the energy content in feeds and two of these are briefly discussed below.

1. Digestibility, measured as a percentage, relates to the portion of food which is not excreted in the faeces and so is available for use by the cow. Digestibility is not a direct measure of energy, but it does indicate overall feed quality. Because cows are able to digest and use more of it, the greater the digestibility, the greater the benefit of that food to the cow.

2. Metabolisable energy (ME) describes the energy in a feed that cows can actually use for their metabolic activities; that is, maintenance, activity, pregnancy, milk production, and gain in body condition. The ME content of a feed can be calculated directly from its digestibility. The ME content of a feed (also called its energy density) is measured as megajoules of metabolisable energy per kilogram of dry matter (MJ ME/kg DM). Intake of ME is expressed in MJ/day.

The higher the energy content of a feed, the more energy is available to the animal. If a feed contains 10 MJ of ME/kg DM, then each kg of dry matter of that feed contains 10 megajoules of ME available for use by the cow. A feed containing 12 MJ of ME/kg DM then has a higher energy content than a feed containing 10 MJ of ME/kg DM.

Describing feed protein

There are various ways to describe dietary protein and these are briefly discussed below.

1. Crude protein (CP). Dietary protein is commonly termed 'crude protein'. This can be misleading, because crude protein percentage (CP%) is not measured directly but is calculated from the amount of nitrogen (N%) in a feed using the formula $CP = N \times 6.25$.

2. Non-protein nitrogen (NPN) is not actually protein; it is simple nitrogen. Rumen microbes use energy to convert NPN to microbial protein for use in the body. In the forage-fed cows, the rumen microbes use NPN with only 80% efficiency (compared to true protein), which reduces its overall value as a protein source. Urea is a source of NPN.

3. Rumen degradable protein (RDP) is any protein in the diet that is broken down (digested) and used by the microbes in the rumen. If enough energy is available in the rumen, some of this RDP will be used to produce microbial protein.

4. Undegradable dietary protein (UDP) is any protein in the diet that is not digested in the rumen. It is digested 'as eaten', further along the gut. That is why UDP is sometimes called 'bypass protein'. The proportion of the protein that is digested in the rumen is called its degradability.

Describing feed fibre

There are various ways to describe dietary fibre and two of these are briefly discussed below.

1. Neutral detergent fibre (NDF) is a measure of all the fibre (the digestible and indigestible parts) and indicates how bulky the feed is. Some of it is digested, and some is excreted. A high NDF might mean lower intake because of the bulk while a lower NDF values lead to higher feed intakes.

2. Acid detergent fibre (ADF) is the poorly digested and indigestible parts of the fibre; i.e. the cellulose and lignin. If the ADF is low, the feed must be very digestible (i.e. high quality).

Crude fibre is another measure of feed fibre, which is becoming outdated but is sometimes quantified when used to describe an alternative measure of feed quality, namely Total Digestible Nutrients (TDN).

Predicting cow performance from nutrient intakes

Cows require nutrients to survive and be productive. It is possible to describe these requirements in terms of dietary energy, protein and fibre. Through knowledge of intakes of energy and the contents of protein and fibre in the total diet, the performance of a dairy cow can be predicted.

Water

In Australia, growing stock under 100 kg require up to 15 L/head/day (or L/hd/d), while stock up to 300 kg require up to 35 L/hd/d. Dry cows require 30–80 L/hd/d and milking cows 70–110 L/hd/d. In the tropics, milking cows require 60–70 L/hd/d for maintenance, plus an extra 4–5 L for each litre of milk produced.

Water requirements rise with air temperature. An increase of 4°C will increase water requirements by 6–7 L/day. High-yielding milking cows can drink 150–200 L water/day during the hot season.

Water intakes are influenced by feed intake, diet (increasing with level of concentrate feeding), humidity, wind speed, water quality (sodium and sulphate levels), and the temperature and pH of the drinking water.

Energy

Cows need energy for maintenance, activity, pregnancy, milk production and for body condition. It is beyond the scope of this manual to describe the detailed calculations required to estimate the energy requirements for dairy cows to achieve set production targets. To formulate any ration for such target production levels, however, such calculations are necessary. These calculations are included in a computer program (MIFC10) freely available from the author (john.moran@dpi.vic.gov.au or jbm95@hotmail.com).

Table 6.1 presents examples of these calculations to determine energy requirements for seven different cows, with varying pregnancy status, and changes in live weight, to produce target yields of milk some with different milk fat and protein concentrations (*Nutrient Requirements of Domesticated Ruminants* 2007). The energy requirements are expressed in MJ of ME per day. The table also includes the proportion of ME required just to maintain the cow. Cows 1, 2 and 3 differ only in milk yield yet the proportion on ME for maintenance increases from 37% to 41% to 47% as milk yields decrease respectively from 30 kg/d to 25 kg/d to 20 kg/d.

Dairy farmers and advisers involved in formulating rations for milking cows should seek professional assistance to ensure the rations they plan to feed provide sufficient energy for cows to achieve production targets.

Protein

The amount of protein a cow needs depends on her size, growth, milk production, and stage of pregnancy. Milk production is the major influence on protein needs. Cows in early lactation require 16–18% CP in their diet and this decreases to 14–16% in mid lactation, to 12–14% in late lactation and 10–12% during the dry period.

Fibre

Cows need a certain amount of fibre in their diet to ensure that the rumen functions properly and to maintain the fat test. Acceptable levels of NDF in the diet are in the range 30–35% DM. Much of this fibre should be sufficiently long (greater than 1.5 cm) to stimulate rumination or cud chewing. This is called long or effective fibre and ideally should make up at least 75% of the NDF in the ration. This can be estimated by sieving the ration to separate different particle sizes physically.

Nutrient requirements of different classes of dairy stock

Most dairy farms have up to seven categories of stock within their herd. The classes of stock and their nutrient requirements, for a typical Friesian herd of 550 kg mature cows, are:

1. Weaned heifers up to 12 months. These require a ration supplying 40–80 MJ of ME/head/day and 15–17% CP.
2. Heifers, from 12 months to point of calving. These require a ration supplying 80–100 MJ of ME/head/day and 13–15% CP.
3. Milking cows in early lactation, and producing on average, 25 L/d of milk during their first 100 days post calving. These require a ration supplying up to 220 MJ of ME/head/day and 16–18% CP.

Table 6.1 Energy requirements of milking cows (in MJ/day of ME) at different stages of lactation and pregnancy status and producing different amounts of milk. It is assumed that milking cows are walking 2 km/d on a flat terrain, thus requiring 4 MJ/d for an activity allowance.

Description	Cow 1	Cow 2	Cow 3	Cow 4	Cow 5	Cow 6	Cow 7
Cow details							
Live weight (kg)	550	550	550	550	550	550	550
Stage of lactation	Early	Early	Early	Mid	Late	Late	Dry
Month of pregnancy	Empty	Empty	Empty	3rd	6th	7th	9th
Milk yield (kg/d)	30	25	20	16	13	10	0
Fat test (%)	3.6	3.6	3.6	3.6	4.0	4.0	0
Protein test (%)	3.2	3.2	3.2	3.2	3.8	3.8	0
Live weight gain/loss (kg/d)	–0.5	–0.5	–0.5	0	0	+0.25	+0.5
Energy requirements (MJ/d)							
Maintenance	75	73	70	67	66	64	62
Activity	4	4	4	4	4	4	0
Pregnancy	0	0	0	0	5	9	33
Milk production	30×4.8 = 144	25×4.8 = 120	20×4.8 = 96	16×4.8 = 77	13×5.3 = 69	10×5.3 = 53	0
Weight gain or loss	–0.5×40 = –20	–0.5×40 = –20	–0.5×40 = –20	0	0	0.25×44 = +11	0.5×61 = +30
Total energy requirements	203	177	150	148	144	141	125
% ME for maintenance	37	41	47	45	46	45	50

4. Milking cows in mid lactation, and producing on average, 18 L/d of milk during their second 100 days of lactation. These require a ration supplying up to 200 MJ of ME/head/day and 14–16% CP.
5. Milking cows in late lactation, and producing on average 15 L/d during their third 100 days of lactation. These require a ration supplying about up to 180 MJ of ME/head/day and 12–14% CP.
6. Dry cows. These require a ration supplying 90–100 MJ of ME/head/day and 11–12% CP.
7. Bulls. These require a ration supplying about 80 MJ of ME/head/day and 12% CP.

Calculations of nutrient requirements of milking cows and nutrients supplied by various combinations of ration ingredients can be quite complex and time-consuming, particularly when there is a choice of feeds. These calculations have also been incorporated into the computer program MIFC10, available free from the author (john.moran@dpi.vic.gov.au or jbm95@hotmail.com).

The lactation cycle

Cows must calve to produce milk and the lactation cycle is the period between one calving and the next. The cycle is split into four phases, the early, mid and late lactation (each of about 100–120 d) and the dry period (which should last as long as 65 d). In an ideal world, cows calve every 12 months.

A number of changes occur in cows as they progress through different stages of lactation. As well as variations in milk production, there are changes in feed intake and body condition, and stage of pregnancy. Figure 6.4 presents the interrelationships between feed intake, milk yield and live weight for a Friesian cow with a 14 month inter calving interval, hence a 360 d lactation.

Following calving, a well-managed cow may start producing 10 kg/d of milk, rise to a peak of 30 kg/d by about seven weeks into lactation then gradually fall to 5 kg/d by the end of lactation. Although her maintenance requirements will hardly vary, she will need more dietary energy and protein as milk production increases then less when production declines. To regain body condition in late lactation she will require additional energy.

If a cow does not conceive, she has no need for additional energy or protein during pregnancy. Once she becomes pregnant she will need extra energy and protein. The calf does not increase its size rapidly until the sixth month, at which time the nutrient requirement becomes significant. The calf doubles its size in the ninth month, so at that stage a considerable amount of feed is needed to sustain its growth.

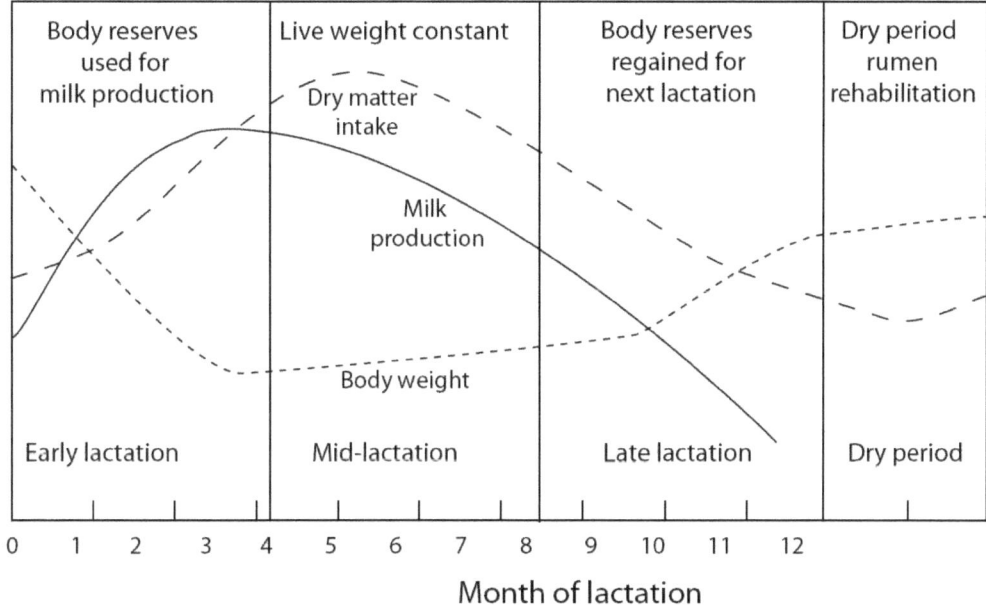

| Body reserves used for milk production | Live weight constant | Body reserves regained for next lactation | Dry period rumen rehabilitation |

Dry matter intake

Milk production

Body weight

| Early lactation | Mid-lactation | Late lactation | Dry period |

0 1 2 3 4 5 6 7 8 9 10 11 12

Month of lactation

Figure 6.4 Dry matter intake, milk yield and live weight changes in a cow during her lactation cycle.

Well-bred dairy cows usually use their own body condition for the first few months after calving, to provide energy in addition to that consumed. The energy released is used to produce milk, allowing them to achieve higher peak production than would be possible from their diet alone. To do this, cows must have sufficient body condition available to lose, and therefore they must have put it on late in the previous lactation or during the dry period.

From calving to peak lactation

Milk yield at the peak of lactation sets up the potential milk production for the year. The full lactation response to extra milk at peak yield varies greatly with feeding management during mid and late lactation. There are a number of obstacles to feeding the herd well in early lactation to maximise the peak. The foremost of these is voluntary food intake.

At calving, appetite is only about 50–70% of the maximum at peak intake. This is because during the dry period, the growing calf takes up space, reducing rumen volume and the density and size of rumen papillae is reduced. After calving, it takes time for the rumen to 'stretch' and the papillae to regrow. It is not until weeks 10–12 that appetite reaches its full potential.

Peak lactation to peak intake

Following peak lactation, cows' appetites gradually increase until they can consume all the nutrients required for production, provided the diet is of high quality. From Figure 6.4, cows tend to maintain weight during this stage of their lactation.

Mid and late lactation

Although energy required for milk production is less demanding during this period because milk production is declining, energy is still important because of pregnancy and the need to build up body condition as an energy reserve for the next lactation. It is generally more efficient to improve the condition of the herd in late lactation rather than in the dry period.

Dry period

Maintaining (or, if required, increasing) body condition during the dry period is the key to ensuring cows have adequate body reserves for early lactation. If cows calve with adequate body reserves, they can cycle within two or three months after calving. If cows calve in poor condition, milk production suffers in early lactation because body reserves are not available to contribute energy. In fact, dietary energy can be channelled towards weight gain rather being made available from the desired weight loss. For this reason, high feeding levels in early lactation cannot make up for poor body condition at calving.

Persistency of milk production throughout lactation

The two major factors determining total lactation yield are peak lactation and the rate of decline from this peak. In temperate dairy systems, total milk yield for a 300 d lactation can be estimated by multiplying peak yield by 200–220.

Hence a cow peaking at 20 kg/d should produce 4000–4400 kg/lactation, while a peak of 30 kg/d equates to a 6000–6600 kg full lactation milk yield. This is based on a rate of decline of 7–8% per month from peak yield; that is, every month the cow produces, on average, 7–8% less of peak yield than in the previous month. Actual values can vary from 3–4% per month in fully fed, lot fed cows to 12% or more per month in very poorly fed cows; for example, during a severe dry season following a good wet season in the tropics.

The higher this number, the faster the rate of decline hence the less milk produced. Persistency is calculated as follows:

$$\text{Persistency (\%)} = \frac{\left[(\text{Peak milk yield}) - (\text{Current milk yield})\right] \div \left[\text{Peak milk yield}\right] \times 100}{\text{Months after peak milk yield}}$$

For example, if cows peaked at 30 kg/d and currently produce 24 kg/d, three months after the peak,

$$\text{Persistency} = \frac{[30 - 24] \div 30}{3} = 7\%$$

If these cows still produced 26 kg/d, their persistency would be 4%, whereas if their milk yield had decreased to 20 kg/d, their persistency would be 11%. So the higher the number, the poorer the lactation persistency. A realistic target persistency over the full lactation would be 7–8% per month.

The rate of decline from peak depends on peak milk yield; nutrient intake following peak yield; body condition at calving; and other factors, such as disease status and climatic stress.

Generally speaking, the higher the milk yield at peak, the lower its persistency in percentage terms. Not only does underfeeding of cows immediately post-calving reduce peak yield, it also has adverse effects on persistency and fertility. Dairy cows have been bred to utilise body reserves for additional milk production, but high rates of live weight loss will delay the onset of oestrus.

How cows respond to supplements

In pasture-based dairy systems, the pasture actually grazed is called the basal forage while all other additional feeds, be they cereal grain, formulated concentrates, hay or silage, are considered supplements. Supplements are then fed to improve or maintain milk production, cow condition or reduce intakes of basal forages, when in short supply.

The impact of supplementary feeding is difficult to assess because the results do not appear immediately as extra milk. In the short term, the response to a particular supplement may be small. But if that supplement is used to 'save' on other feeds, for example, until the pastures become more readily available or until a lower-priced supplement (such as a crop residue) can be used, then it may still be a good management decision and ultimately profitable.

The factors affecting responses to supplementary feeding are numerous (see Figure 6.5) and their interactions are complex.

Substitution for basal forage

Dairy cows may eat less pasture when supplements are fed. The term 'substitution' describes the extent to which a supplement replaces the pasture which would otherwise have been eaten had the supplement not been offered. The substitution rate is the reduction in pasture DM intake per kg DM of the supplement offered. A

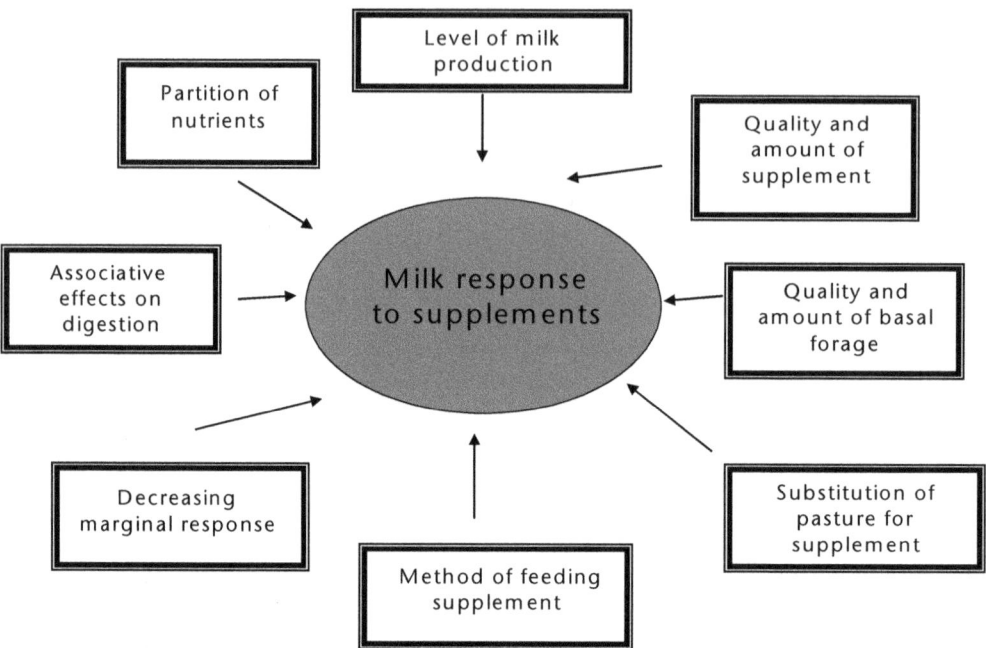

Figure 6.5 Some of the key factors influencing milk responses to supplementary feeding.

substitution rate of 0.25 then means that for every kg DM of supplement eaten, pasture intake will fall by 0.25 kg DM.

The major factor influencing substitution rate is the pasture intake achieved ·with no supplements fed. If the supplements are fed when the pasture has not been well-utilised, it is likely that there will be little response in milk production and much of the pasture will be wasted.

Substitution rates vary with:

- Amount of pasture offered each day – as more pasture is offered, substitution is likely to increase
- Intake limit of the cow – the closer the cow is to the limit of her feed intake when a supplement is offered, the greater the substitution.
- Pasture quality – if the pasture is the same or poorer quality than the supplement, then the supplement may be preferentially eaten.
- Type of supplement – substitution is generally greater with roughage (hay, silage) compared to concentrate supplements. This reflects the volume the supplements occupy in the rumen and how quickly they are digested to make room for more feed (i.e. roughages are bulky and are digested slowly).
- The expected degree of substitution attributed to different supplements may not be an issue until supplement feeding levels exceed 5 kg DM/cow/d and pasture allowances are moderate (Stockdale *et al.* 1997).

- Balance of the diet as a result of feeding the supplement – if the supplement corrects a dietary imbalance, less substitution occurs and intakes may increase. If an imbalance is made worse, total intake may be reduced.

The impact of substitution on the profitability of supplementary feeding depends on the relative costs of the grazed pasture and the supplements. If the cost of pasture DM actually consumed by the grazing cow is lower than the cost of the supplement DM, then this will reduce any monetary benefits derived by any additional milk produced. This is usually the case with pasture-based dairy farming except when the input costs for pasture production increase markedly, as they have during the recent drought in southern Australia. Assessing the profitability of intensifying dairy farming is a lot more complex than this as higher yielding cows are generally more efficient converters of feed into milk (see Figure 6.5). Such improved feed conversion efficiencies are discussed in Chapters 11 and 13.

Decreasing marginal responses

As the intake of energy increases, the amount of extra milk produced from each extra unit of energy can vary considerably. Initially increasing amounts of energy may lead to more milk per unit energy, but eventually milk responses will level off and can even decrease. In other words, the marginal, or additional, milk response eventually decreases as the level of supplement intake increases.

The major reason for this decreasing marginal milk response is that, with successive increments of feed energy, the cow increasingly partitions nutrients from milk production towards body tissue deposition as milk production approaches the cow's genetic potential. In addition, the stage of lactation has an influence on how much of the supplement's nutrients 'go into the bucket' and how much 'go on the back'. Cows in early lactation tend to lose weight to divert additional nutrients towards milk while those in late lactation tend to repartition nutrients to replace previously lost body reserves.

A second reason for declining marginal responses is that utilisation of one feed type can change with increasing intake of a second feed type, which is known as an associative effect. Efficient digestion of forages requires small variations in rumen pH so that adequate fibre digesting microbes can survive and thrive in the rumen. If the supplement is cereal grain or some other high starch concentrates, the proportion of these microbes will decrease as more starch-digesting microbes propagate as a result of a lower pH. Consequently the digestion of the forage can decrease with increasing intakes of such concentrates. Additional starch excretion may also occur, further reducing feed utilisation. This can be particularly important when feeding high levels of supplements rich in fermentable carbohydrates during milking, as rumen pH can decrease, markedly reducing fibre digestion.

The degree to which associative effects can influence milk responses depends on many factors, the most important ones being:

- the type of pasture, conserved fodder and concentrates being fed;
- the amounts of these feeds being consumed;
- their nutritive characteristics (composition, physical characteristics and rates of fermentation); and
- management practices that affect the pattern of feed intake, such as pasture allocation and method and frequency of feeding supplements.

Supplementary feeding usually results in higher total feed intakes. Increasing intakes are the result of decreased times that consumed feed spends in the rumen where it is exposed to microbial breakdown. If less of that feed is digested and the nutrients absorbed into the bloodstream or pass down the digestive tract, less dietary energy becomes available for use by the animal. The cow partly compensates for this through decreased losses of energy via methane and urine with increasing feed intake. Although this may not be important unless total feed intakes increase through supplementation dramatically, it can contribute to declining marginal milk responses to supplements.

The principle of decreasing marginal responses means that any mathematical model predicting milk responses to changes in the type and/or amount of basal forage or supplement offered the milking herd or the intake of specific feed nutrients, should incorporate curvilinear rather than linear statistical relationships. This is rarely the case with dairy (or in fact for any biological) predictive models. Granted the biological relationship may essentially be linear for the initial few levels of nutrient increments; however, for intensive dairy systems, the assumption of linearity throughout the response curve can, and does, overpredict actual milk responses; this is discussed further in Chapter 8. For this reason, relying purely and entirely on the theoretical results, rather than also using experience hence some degree of subjectivity, can lead to poor decision making when developing feeding systems for intensively managed grazing dairy cows.

Just because supplemental nutrients may not be expressed through extra milk volume does not negate the economic benefits of supplementary feeding. Higher yielding cows are generally more efficient energetically, as discussed in Chapter 11. They can also generate additional income in the medium to long term through higher yields of milk solids (which can occur when supplemental energy overcomes an energy deficiency in grazing cows), improved live weight gain (either as muscle to increase potential mature live weight or to partitioning more nutrients (as fatty tissue) into body reserves or foetal growth and even greater resistance to disease agents (through improved animal immunity and vigour). In addition unit operating and fixed costs can be diluted over higher farm milk outputs.

An example of marginal milk responses

The best way to understand the concept of decreasing marginal milk responses more fully is through an actual example from a study on supplementary feeding. The following graph originates from a study undertaken with grazing cows in mid lactation that were supplemented with increasing levels of crushed barley while grazing summer perennial pastures (Gibb *et al.* 2006). The cows were fed the barley twice daily at milking. Figure 6.6 presents the milk responses in terms of g/d of milk solids (that is, fat and protein). Because farmers in Victoria are paid on daily production of milk fat and protein, rather than volume of milk, they are being encouraged to use kg/d of milk solids rather than L/d of milk when quantifying milk yields. With milk solids averaging 7.5%, 75 g/d of milk solids is equivalent to 1.0 L/d of milk.

Figure 6.6 presents two sets of milk responses to increasing barley intakes. The *average milk response* (AMR) is calculated from the quantity of extra milk solids produced by total barley intake whereas the *marginal milk response* (MMR) is the extra milk solids produced by each increment of barley supplement. The AMR and MMR are obviously the same (63 g/kg) at the first increment of barley feeding

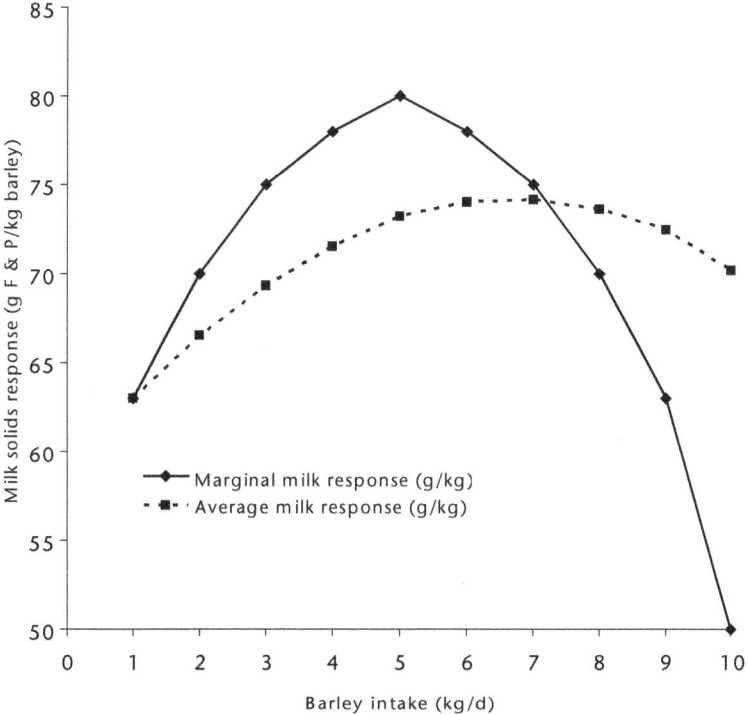

Figure 6.6 Changes in average and marginal milk responses in cows grazing summer perennial pastures in northern Victoria and supplemented with increasing levels of concentrates (Gibb *et al.* 2006).

(1 kg/d). From then on, however, they are different. The highest MMR (80 g/kg) occurs at 5 kg/d barley, whereas the highest AMR (74 g/kg) occurs at 7 kg/d of barley.

From a knowledge of the supplement cost and the milk return, more informed decisions can then be made on the most economic level to feed particular supplements. For example, if barley costs $250/t (25 cents/kg) and milk solids returned $3.60/kg, the break-even price milk response (when cost equalled return) is 70 g/kg. This occurred at 10 kg/d of barley using AMR but only at 8 kg/d of barley using MMR.

Clearly, if using AMR to decide on the optimum level of grain feeding, farmers would feed 7 kg/d whereas using MMR, the optimum is 5 kg/d of barley. Furthermore, from the point of view of increasing milk returns, there seems little point in feeding more than 7 or 8 kg/d of barley, because of the very poor marginal milk responses.

The data presented in Figure 6.6 is derived from one extreme system in which cows were supplemented twice daily with a high starch cereal grain, but with no additives to stabilise rumen pH. The key role of feedpads is to allow the feeding of more formulated supplements (containing forages as well as concentrates) throughout the day. In this case, both AMR and MMR could differ markedly from

Figure 6.7 Round bale silage stretch wrapped in line.

those in Figure 6.6. The diminishing marginal responses with PMRs are likely to be less dramatic than those in Figure 6.6 because PMRs, being based on forages as well as concentrates and incorporating additives (see Chapter 7), would more likely maintain rumen pH hence reduce any associative effects on digestion of the forage component of the total diet (Leddin and Armstrong 2009).

As previously mentioned, milking cows can show financial benefits in addition to higher milk returns from supplementary feeding. But as these are often difficult to quantify, the decision on the best level to feed such a supplement is not easy.

Feeding formulated supplements

When grazing cows are fed additional supplements, these can be in the form of 'slugs' of concentrates in the milking shed (as in Figure 6.6), conserved forages (as in Figure 6.7) or rations formulated to provide a diversity of feeds. Supplements fed to grazing cows while on a feedpad are likely to be mixtures of various types, such as cereal grains or high starch (or other rapidly fermented carbohydrate) concentrates, fibrous agro-industrial by-products (for example citrus pulp or brewers grain) and conserved forages (namely hay or silage). Such a ration, called a partial mixed ration (see Chapter 7) is likely to lead to a more stable rumen pH hence a smaller associative effect than would be the case with cereal grain supplements fed during milking.

Another factor decreasing milk responses is the often-incorrect assumption that all of the supplement is actually consumed. Rarely is there nil wastage, particularly if the supplement is a roughage. The hidden costs of feed wastage are discussed in Chapter 11.

The major difficulty when predicting milk responses to supplementation, even if substitution rates are known, is the lack of information on the relative importance of the above factors. Without such knowledge, dairy advisers can only, probably incorrectly, assume additive effects when feeding a mixture of various feed types, which would tend to overestimate such milk responses. This is particularly the case when there are marked differences between basal roughages and supplement type, and large amounts (say, 6 kg DM/cow/d or more) of supplement are fed.

Immediate and delayed milk responses

Responses to supplementary feeding have both immediate and delayed components. Some of the supplement goes immediately to milk production while some goes to body reserves, which contribute to milk production at a later stage (usually during the following lactation) when these reserves are mobilised.

Most reported milk responses are only the immediate ones because the delayed responses, derived from mobilised body reserves, are rarely monitored.

The guidelines for grazing cows in temperate dairy regions (Target 10, 2005) are:

- In early lactation, the average immediate response to feeding concentrates containing 12 MJ/kg DM of ME is 0.6 kg of milk per kg of supplement DM, ranging from 0.2–1.0 kg.
- In mid-lactation to late lactation, the average immediate response is 0.5 kg of milk per kg of supplement DM, ranging from 0.2–0.8 kg.

Theoretically a concentrate containing 12 MJ of ME/kg DM could produce:

- an additional 2.2 kg of milk per kg supplement DM (for milk containing 4.0% fat and 3.8% protein and requiring 5.5 MJ/kg)
- an additional 2.4 kg of milk per kg supplement DM (for milk containing 3.8% fat and 2.6% protein and requiring 5.0 MJ/kg)

There is a large discrepancy between the theoretical responses (2.2–2.4 kg milk/kg supplement DM) and those actually recorded during early lactation, in controlled studies (0.2–1.0 kg milk/kg supplement DM). This is the result of the many contributory factors reducing potential milk responses as discussed in this chapter. It also highlights the difficulties in just predicting milk responses, let alone economic benefits to supplementary feeding.

To date, virtually all studies have been undertaken with energy and/or protein rich concentrates. There is an urgent need for corresponding studies in which grazing cows are supplemented with partial mixed rations containing combinations of concentrates and forages.

7

Partial mixed rations

This chapter introduces the concept of partial mixed rations and the range of potential ingredients and discusses ways to troubleshoot feeding problems.

The main points in this chapter:

- Formulated rations can be total mixed rations (TMR) or partial mixed rations (PMR) depending on whether the stock have access to grazed pasture.
- The nutritive value of various supplements (whether fed for energy, protein or fibre) is very variable and such supplements can be classified on their supply of dietary energy and protein.
- In addition to cereal gains, farmers have access to a wide range of agro-industrial by-products. They should take precautions to ensure these do not contain undesirable chemical residues which can affect feed safety and animal performance adversely. Different by-products can be sourced as liquids, very wet or quite dry, and this can create specific problems with storage and feeding out.
- Many simple observations can be used when troubleshooting feeding problems. These include sudden changes in appetite, manure consistency, milk yield and milk composition. Feed toxicities and moulds can limit animal performance while the feeding of urea requires specific guidelines.
- Metabolic disorders can also be brought about by feeding unbalanced diets. Unless the pH in the rumen is maintained within strict limits, feed digestion and cow performance will suffer.
- Lactic acidosis can be a common metabolic problem resulting from feeding excess grain without taking precautions to maintain rumen pH. There are a variety of feed additives which can help overcome such problems.

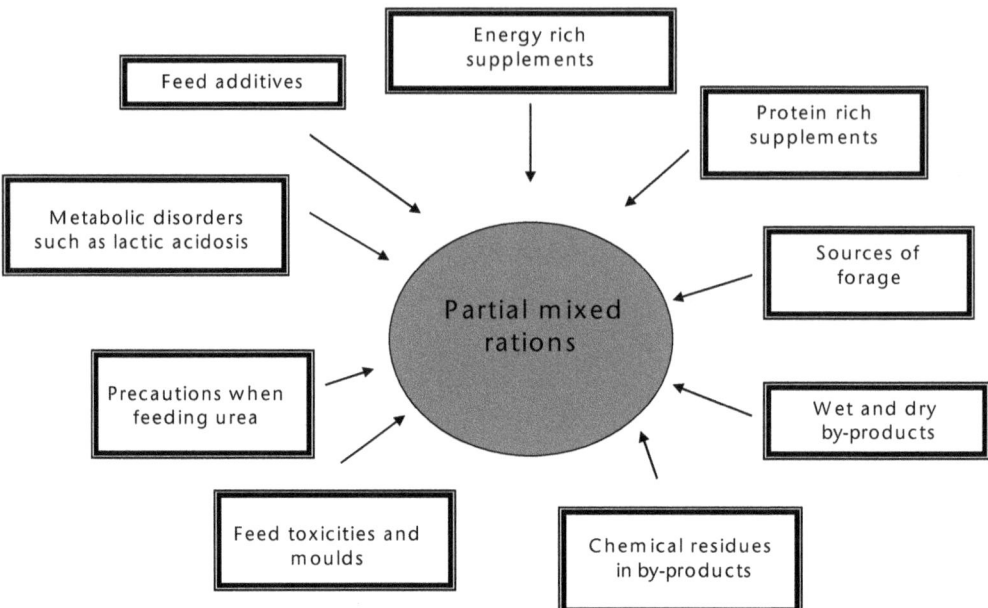

Figure 7.1 Some of the key considerations when formulating partial mixed rations.

Just as feedpads can vary from a series of hay rings in a sacrifice paddock to a free stall shed with all its associated facilities, the formulation and feeding out of supplementary feeds also ranges from quite simple or extremely complex.

Hay or silage fed out in the paddock, or for that matter on a feedpad, require little knowledge on ration formulation, because only one form of supplement is fed to the grazing cows, although they usually receive concentrates in the milking shed. When deciding to make better utilisation of the feedpad, however, a new skill, namely formulating rations, is required. This chapter complements Chapter 6 in that it applies practical applications to the previously discussed theories on dairy cow nutrition. These include using feed nutrient content to select the most appropriate ration ingredients, precautions to take when feeding agro-industrial by-products and how to recognise and treat feeding problems in partial mixed rations.

Formulating rations for dairy stock on a feedpad

Total and partial mixed rations

There are two terms that are frequently used in dairy nutrition circles which are important to farmers feeding stock on feedpads.

The first term is *total mixed rations* or *TMR*, which is the fully formulated ration fed in a feedlot or feedpad to stock with no access to grazed pasture. The stock are then completely dependent on the farmer for their full supply of feed nutrients (energy, protein, fibre, minerals and vitamins). TMRs are formulated from the knowledge of the cow's nutrient requirements and the nutrients supplied by the ration ingredients. They are also known as complete diets.

The second term is *partial mixed rations* or *PMR,* which is the formulated ration supplied to grazing stock while on the feedpad (see Figure 7.1). Assuming they have already consumed some pasture, this ration must take into account the feed nutrients consumed in the paddock and the milking shed. The major skill for the farmer is then to try and predict what the cows have consumed while grazing and just supplement this with sufficient feed nutrients to satisfy their requirements to achieve whatever production target the farmer has set for his cows. That is not always easy because of the many factors that influence both the amount of forage and its quality as selected by grazing stock. These include the time spent grazing; time of day at pasture; the availability and quality of pasture in the paddock; the cows' desire to graze, or appetite; any climatic stress that may reduce their appetite; water supplies while at pasture; and any other influences on appetite, such as health, pregnancy, and social interactions with other cows in the herd.

The principles of formulating feedpad rations

As grazed pastures are the basis of most dairy production systems in Australia, they should be managed to produce the most economical forage sources. This does not necessarily equate with their maximum forage yields because quality is usually lost when farmers chase high pasture yields. The feedpad system allows milk production to be maintained in cows when pasture supplies are limited. Low milk yields can be energetically inefficient because too much of the consumed nutrients are devoted to just maintaining the cow rather than being available to produce milk. This was discussed in the previous chapter (see Table 6.1).

As mentioned in Chapter 6, PMRs are likely to lead to lower associative effects than cereal grain-based supplements because of the inclusion of forages. Once a target milk yield is set, the next step is to estimate the amount of the required daily feed nutrients that can be supplied by the grazed pasture. This will provide some indication of how much energy and/or protein should be supplied on the feedpad or in the milking shed. With a TMR, however, bail-feeding in the milking shed is not really necessary.

The total daily nutrients are then supplied by both grazed pasture and PMR. The energy supplied on the feedpad can range from less than 5% to 95% or more, depending on the availability of pastures for grazing. This will vary with season, stocking rate, provision of concentrates during milking and other factors decided on by the farmer. Such discussions are outside the scope of this book.

The provision of energy in the PMR should be to balance grazed pasture supplies. The provision of protein should be to provide a daily content of protein in the total ration (PMR and grazed pasture) ranging from 12–16%, depending on stage of lactation. To maintain normal rumen function, the daily fibre supplies should provide at least 35% NDF in the total ration DM for all classes of dairy stock.

Nutritive value of ration ingredients

Most dairy cattle nutritional reference books contain tables of nutritive value of feeds and supplements. The following two tables provide a summary of some of the common forages (Table 7.1) and supplements (Table 7.2) fed to dairy cows around Australia. The ranges take into account difference in the agronomic growing conditions and stage of maturity in fresh and conserved forages and cereal grains. They also take into account the diversity of processing of agro-industrial by-products prior to their purchase as supplements. The values in Table 7.1 cover typical ranges throughout the year while those in Table 7.2 cover the extremes found in feeds sourced by dairy farmers around Australia (Jenkin and Wales 2007). Table 7.3 classifies these feeds on the basis of their mean CP and ME contents from Tables 7.1 and 7.2.

The data in Table 7.1 were derived from laboratory analyses of sampled forages. Grazing stock, however, have the ability to select specific parts of the plant, such as the leaves and fine stalks. Therefore the nutritive value of the pasture actually selected by the grazing stock is likely to be higher than that in the pasture on offer. The term 'selection differential' is used to quantify the ability of grazing stock to select the better quality pasture.

Typical selection differentials for grazed pastures in Victoria (Doyle *et al.* 2000) expressed as % of the value in pasture harvested to ground level are:

- Energy: 105–110%, equivalent to an increase of 0.5 to 1.0 MJ/kg DM for a pasture containing 10.0 MJ of ME/kg DM.

Table 7.1   Ranges in nutritive values of some common Australian forages.

Forage	DM (%)	ME (MJ/kg DM)	CP (%)	NDF (%)
Grasses				
Ryegrass	14–20	9–13	14–20	30–50
Paspalum	14–25	9–12	14–18	30–60
Forage sorghum	13–15	7–9	11–15	65–75
Native pasture	25–35	7–9	8–10	60–70
Clovers				
White clover	12–18	10–13	20–30	25–50
Subclover	12–18	10–13	20–30	25–50

DM, Dry matter; ME, Metabolisable energy; CP, Crude protein; NDF, Neutral detergent fibre.

Table 7.2 Mean (and range) in nutritive value of some common Australian supplements.

Feed	DM (%)	ME (MJ/kg DM)	CP (%)	NDF (%)
Energy supplements				
Barley grain	88 (85–90)	12 (9–13)	11 (6–19)	20 (14–36)
Maize grain	88 (85–90)	14 (12–15)	10 (7–22)	9 (6–13)
Oats grain	88 (85–90)	10 (6–14)	9 (4–15)	29 (22–34)
Triticale grain	88 (85–90)	13 (12–14)	11 (7–19)	15 (11–23)
Sorghum grain	88 (85–90)	13 (12–14)	11 (10–13)	7 (6–8)
Wheat grain	88 (85–90)	13 (10–14)	13 (7–23)	12 (8–24)
Citrus pulp	14 (10–17)	13 (10–14)	9 (6–12)	25 (18–34)
Tapioca pellets	86	13	3	5
Brewers grain	28 (14–60)	11 (8–14)	22 (10–29)	55 (42–62)
Protein supplements				
Cottonseed meal	90 (88–92)	11 (10–13)	43 (40–48)	31 (21–39)
Canola meal	90 (88–92)	12 (10–16)	37 (27–42)	30 (29–35)
Copra meal	90	12	20	52
Lupins	92 (90–95)	13 (11–15)	32 (21–43)	23 (9–29)
Palm kernel extract	94 (92–96)	11 (9–12)	16 (15–16)	65 (55–74)
Dried distillers grain	92 (90–94)	14 (12–15)	27 (16–43)	29 (18–39)
Grape marc	55 (20–94)	6 (2–12)	12 (5–18)	48 (20–61)
Forage supplements				
Barley silage	39 (25–60)	9 (5–11)	11 (5–23)	61 (44–69)
Barley hay	87 (85–90)	9 (4–11)	8 (1–15)	58 (42–87)
Barley straw	89 (88–91)	6 (2–8)	3 (1–30)	76 (55–87)
Canola silage	47 (25–76)	10 (7–12)	18 (10–26)	38 (26–52)
Canola hay	85 (61–93)	10 (7–13)	17 (9–27)	38 (25–53)
Clover silage	42 (30–60)	10 (8–11)	19 (12–27)	46 (39–56)
Clover hay	87 (85–90)	9 (6–11)	18 (6–26)	47 (33–72)
Grass silage	43 (20–70)	9 (5–12)	13 (5–27)	58 (40–78)
Grass hay	86 (84–90)	8 (5–10)	8 (1–18)	67 (43–83)
Lucerne silage	50 (20–80)	9 (4–11)	20 (15–32)	46 (27–64)
Lucerne hay	88 (85–90)	9 (5–11)	19 (6–30)	45 (30–67)
Wheat silage	45 (30–60)	9 (5–11)	10 (6–16)	56 (47–63)

Table 7.2 continued

Feed	DM (%)	ME (MJ/kg DM)	CP (%)	NDF (%)
Wheat hay	87 (85–90)	9 (5–11)	8 (1–17)	53 (37–79)
Wheat straw	92 (90–94)	5 (4–9)	3 (1–9)	73 (54–86)
Oaten silage	41 (20–70)	9 (6–11)	10 (4–19)	60 (40–75)
Oaten hay	89 (85–92)	8 (4–11)	7 (1–19)	60 (41–84)
Oaten straw	89 (87–91)	6 (4–10)	3 (1–16)	73 (54–79)
Sugarcane top hay	93 (90–95)	7 (3–9)	6 (3–10)	67 (57–77)
Rice straw	85 (65–94)	7 (5–9)	4 (2–5)	63 (53–68)
Maize silage	31 (15–40)	10 (5–13)	8 (3–17)	48 (36–67)
Almond hulls	90 (88–92)	10 (8–10)	5 (4–6)	35 (30–45)
Whole cottonseed	94 (93–96)	13 (11–15)	23 (15–28)	55 (43–72)

DM, Dry matter; ME, Metabolisable energy; CP, Crude protein; NDF, Neutral detergent fibre.

- Protein: 120–130%, equivalent to an increase of 3–5% units for a pasture containing 16% CP.
- Fibre: 80–90%, equivalent to a decrease of 4–8% units for a pasture containing 40% NDF.

Table 7.3 provides a ready reckoner to consider when seeking feeds to balance cow diets which may be low in dietary energy and/or protein.

Agro-industrial by-products

The use of by-products and alternative feeds has increased substantially in recent years. In the past, they were used to supplement low quality roughages but with

Table 7.3 Classification of supplements according to their average energy and protein contents.

Energy/protein classification	Poor energy (<8 MJ/kg DM of ME)	Moderate energy (8–10 MJ/kg DM of ME)	Good energy (>10 MJ/kg DM of ME)
Poor protein (<10% CP)	All cereal straws Sugarcane top hay	Cereal hays Grass hay Wheat silage Oat silage Almond hulls	Oat grain Maize silage Citrus pulp Tapioca pellets
Moderate protein (10–16% CP)	Grape marc	Pasture silage Cereal silages	Most cereal grains Palm kernel extract
Good protein (>16% CP)	Urea (0 MJ/kg DM of ME 281% CP)	Lucerne hay/silage Clover hay/silage Canola hay/silage	Lupins Brewers grain Whole cottonseed Dried distillers grain All protein meals

ME, Metabolisable energy; CP, Crude protein.

more widespread use of mixer wagons and a better understanding of their nutritive value, they are more commonly used in full production rations. Although poultry litter and animal by-products have previously been used protein sources, it is now illegal to include them in cattle rations due to animal and human health concerns. Another factor to consider is whether the by-products are likely to contain any genetically modified (GM) material, as some milk processors have commercial restrictions on their use.

Because their DM and nutrient contents are often very variable, these should be monitored by regular testing. Where their composition varies widely, regular adjustments to the ration are necessary or else animal performance may suffer.

Before using any by-product, answers to the following questions should be sought:

- Does it contain potentially toxic or banned compounds, such as chemical residues and anti-nutritional factors?
- What and how variable is its nutrient value; that is, its DM, energy, protein and fibre contents?
- Is the material palatable and acceptable to the stock?
- Is it likely to lead to tainting of milk?
- Are there any health problems associated with its feeding? As well as toxicities, this includes poisonings and physical obstructions (in the gut or the oesophagus).
- Does it contain extraneous material such as metals, plastics or other physical contaminants?
- How much material is available, when and where? How seasonal are the supplies?
- What is its true cost per kg DM 'down the cow's throat', when transport, bulk density and DM content and losses during storage and feeding out are all taken into account?
- How can it successfully be stored?
- What extra handling and storage facilities are needed?

Chemical residues

Some by-products may be contaminated by chemical residues from pesticides or other chemical treatment during processing. Unacceptable residues may still be present in the waste plant material after processing or in crop residues when fed out. Residue risks can increase by some chemicals being concentrated in the crop waste fraction. A harvest-withholding period (as specified on most chemical labels) does not guarantee that other parts of the crop, such as stubble and trash, are suitable for stock-feed. Materials such as grape marc, pomace, citrus peel and vegetable skins and outer layers of leafy vegetables often have higher residue levels than the commodity from which they are derived. Some chemicals registered for use on fruit and vegetables are not registered for use on stock feeds or directly on livestock.

Maximum residue limits (MRL) are set for a range of chemicals in food commodities and animal feeds. MRLs must not be exceeded in food products such as milk and meat. Where no MRL is set, a food product may be condemned if there is any residue whatsoever because any detectable levels of that chemical would breach food standards and thus it cannot be used for human consumption. In addition, MRLs acceptable in Australia may not meet standards of our overseas trading partners.

Withholding periods (WHP) for any chemical use are designed to ensure food products do not have residues that exceed Australian MRLs, provided the chemical is used in accordance with their label directions. However, bear in mind that specific WHPs for crop chemicals may differ depending on whether the crop is harvested, grazed or conserved for fodder; failed crops may be harvested before the WHP has expired and so will remain contaminated; and some crops, not normally used for stock feeds, may have had chemicals applied on them which do not carry any specific label warnings.

In some states it is an offence to feed waste from treated crop contrary to label directions. So farmers should check with their appropriate state agencies. There is further risk from by-products grown on contaminated land. For example, organochloride residues have been found when animals were fed by-products such as sugarcane tops or vegetable wastes, harvested from contaminated land.

Before buying or accepting the waste material, representative samples should be analysed for pesticide residues by an accredited laboratory, requesting specific analyses. With increasing demands from end-users, some by-product suppliers routinely test their products for chemical residues. A signed Vendor Declaration Form (VDF) should be sought to provide information on the chemical treatment history of the product in question and to verify its chemical residue status.

Buyers of by-products that may or may not contain chemical residues should take precautions by also recording the date when the by-product was received; type of feed; source of supply; analyses carried out; which stock received was actually fed to live stock; dates when the by-products was fed; and length of feeding period. Farmers should also store a sample of the by-product for 12 months.

Types of by-products

By-products generally include discarded or secondary fraction co-products derived from the manufacturing of foods, drinks oil, fuel etc. Generally demand exceeds supply so a long-term view should be taken as there are often no spot markets available. As the suppliers or merchants of these feeds often do not have direct control of their production, it is imperative that purchasers seek an indication of their reliability of supply. Their long-term availability depends on crop forecasts, rainfall and in the case of biofuels, government legislation. Extra storage facilities will allow additional volume to be held on-farm to cover any shortfalls in their supply. Their economic potential is significantly diminished without consistent and predictable nutritional analyses.

Commonly used by-products include fruit and vegetable by-products such as citrus pulp, processed potatoes, almond hulls or grape marc; brewing or distilling by-products such as brewers grain, distillers grains or malt culms; oilseed by-products such as canola, cottonseed, soy bean or sunflower meals; legume offal such as bean, pea or lentil kibble and bean, pea, lentil pollard and husks or hulls; liquid feeds such as sugarcane molasses, citrus molasses, ME 14 (a by-product of cheese making).

By-products can be categorised into wet or dry as described below.

By-products from abattoirs can no longer to be fed directly to cattle or sheep. Meat and bone meal and tallow used to be valuable livestock supplements but since the outbreak of 'mad cow disease', their use has been banned in Australia and many other countries.

Wet by-products

Many by-products have low DM contents making long-term storage difficult. With good silage making practices, however, most by-products can be successfully ensiled. The most important goal is to re-establish an anaerobic environment as quickly as possible and to promote lactic acid fermentation. A silage inoculant may assist with this process. The by-product stack must be effectively sealed with plastic sheeting to prevent large storage and quality losses.

Their high moisture contents make transport costly so they should be costed against alternative feeds on a delivered basis (c/MJ of ME or $/kg CP). Even when supplied free ex-factory, any transport and handling charges may result in an expensive feed. Carting water in high moisture products is very expensive! In these circumstances, the factory may need to pay for their disposal.

Regulations govern the transport of some by-products, such as grapes and vines (for phylloxera) and fruit (for fruit fly). The actual transport routes should be planned to minimise complaints, particularly in cities and towns. Loose products must be covered to avoid spillage. In some instances, curfews limit site access. Storage bunkers should provide for vehicle movement at least five metres around them. The farmer is obviously responsible for collecting and disposing of spills, product residues, waste plastic and any rainfall runoff from the storage area.

Most by-products are primarily sources of energy and have low protein content although some, such as brewers grain are also protein sources. A range of by-products can be ensiled; for example, apple pomace; brewers grain; citrus pulp; distillers grain; grape marc; pineapple pulp; potato waste; tomato waste; sweet corn stover; sweet corn trash, from processing plants; vegetable residues, such as asparagus butts; fresh fruits and vegetables, such as potatoes, carrots and bananas; and time-expired fruit and vegetables.

Liquid by-products, requiring tanks for storage and mixer wagons for feeding out, include molasses (sometimes fortified with additional nutrients); residues from cheese making, such as whey or ME14, and distillers grain syrup

The chemical analyses of some of these have been presented in Table 7.2. Further details are provided by Griffiths *et al.* (2004).

Dry by-products

Dry by-products are easier to transport and store but must be well-protected from the environment. If they become wet, not only will their quality deteriorate, but moulds and other toxic microbes can grow and contaminate them.

Many of these by-products are sources of protein as well as energy. The range of dry concentrates fed to dairy stock include bakery waste; corn hominy; corn gluten meal; cottonseed meal and whole cottonseed; canola meal; copra meal; dried distillers grain; fish meal; linseed meal; malt culms; millrun (wheat); palm kernel extract (imported from Asia); rice bran and pollard; soy bean meal; sunflower meal; tapioca pellets (imported from Asia); time-expired or unsaleable food products such as stale bread, bakery products or confectionary seconds; and wheat bran and pollard.

High fibre by-products, excluding cereal straws and other stubbles (such as maize stover) include chickpea gradings; dried brewers grain; dried grape marc; hulls, such as almond, cottonseed, oat, rice and peanut shells; sugarcane bagasse; and whole cottonseed.

Other very high fibre, and poor energy/protein, sources that can be fed to dairy stock include cardboard; corrugated cardboard boxes; newspaper; and sawdust.

The chemical analyses of some of the above have been presented in Table 7.2 while further details are provided by Dairy Australia (2007a and 2008a) and Mickan (2009).

Troubleshooting potential feeding problems

There are many simple observations farmers can use to highlight potential problems when formulating and feeding out rations on the feedpad. Such quick checks include:

- *The physical appearance and smell* of the concentrate and forage components of the PMR.
- *Cow visual appearance* – if the cows are looking poor, with dull sunken eyes, scruffy coats or hunched back, feeding management should be reviewed. The condition (appearance and cleanliness) of the hair coat is also a good guide to adequacy of feeding management.
- *Rumination* – what proportion of the cows are chewing their cud (ruminating) while at rest? If it is less than 50%, then diet fibre levels should be checked.

Figure 7.2 Commodities can be stored in simple bunkers using hay bales for walls (top) or in more sophisticated cement bunkers (bottom).

- *Mobility of the legs and feet* – this can be assessed easily using the locomotion score described in Table 9.1 in the chapter on cow management.
- *Respiration rate, coughing and nasal discharges* – as well as highlighting feeding problems, this can indicate signs of ill health or heat stress. This is discussed in more detail in Chapter 9.

- *Body condition* – rapid or sustained loss in body condition in otherwise healthy cows can indicate poor energy intakes.
- *Sudden changes in appetite* as is evident from concentrate residues in the milking parlour or excess post grazed residues in the paddock.
- *Sudden changes in daily milk yield* – drops in the milk vat can occur overnight due to sudden climatic changes, but if they last over a number of days, then energy could be lacking. Drops of 1–2 L/cow/day over two to four days should be investigated. They could be due to insufficient access to clean drinking water as well as shortages of grazed pasture and/or supplements.
- *Sudden changes in milk fat and protein tests* – compare current tests with those the same time last year, using 10-day averages. If milk fat levels are more than 0.2% below the same time last year (or even the previous 10-day period), check that the diet has adequate fibre. If milk protein is 0.2% below the same time last year, consider increasing energy intakes through additional pasture allocation or supplements. If milk protein levels do not respond within five days, seek additional nutritional advice. When comparing this year and last year, be sure to take into account any changes in herd calving pattern and milk yield. Changes in milk composition are discussed further below.
- *Changes in manure consistency, colour and content* – manure can be scored for its consistency using Table 7.4, which also suggests any action required. Score at least 25 manure pats as you walk among the herd in the paddock and calculate the average. It is important to assess the range in consistency, because if more than 20% of the pats are one score or greater (or less) than the average, this may also indicate a nutritional imbalance or management problem.
- *Cereal grain residues in manure pats* – when whole cereal grains are purchased for on-farm processing, manure should be checked for whole grains. Only take note of intact grains, those with starch in the grain and disregard grain husks. Excessive amounts of intact whole grains in manure indicate inadequate digestion. The likely causes are the grain has not been processed or crushed adequately or the diet may be deficient in fibre.

Keep in mind that a sudden change in one of these quick checks may be due to a temporary fluctuation in nutrition. Provided that the check quickly returns to normal, cow performance may not be adversely affected. It is important to take action when a quick check remains abnormal for several consecutive days and/or several quick checks become abnormal at the same time.

Feed toxicities

Incorporating often different or unusual feeds, into PMRs can result in cows reacting adversely, due to the presence of feed toxicities, such as Prussic acid (cyanide) in some varieties of forage sorghums, when harvested or grazed as immature crops, or when drought-stressed; Mimosine in some varieties of *Leucaena;* Gossypol in whole cottonseed and cottonseed meal; tannins in banana stems and leaves and mango seed kernel; and Trypsin inhibitor in soy bean products.

Mouldy silage can contain bacteria such as *Listeria*, which can cause abortions in dairy cows; it can also pass through the milk and infect humans. Noxious fungi in silage can lead to pneumonia and abortions in cows.

The high humidity and temperature of tropical environments can encourage the growth of many contaminating microbes. For example, mycotoxins and aflotoxins such as *Fusarium* and *Aspergillus* can grow on moist cereal grains and by-products. Feed ingredients, of both plant and animal origin, can be contaminated with *Salmonella,* which can cause disease and death, particularly in young calves.

The effects of mould on animal health

The effects of mould on animal health and production are not clear-cut.

Mycotoxins can have a very negative yet sub-clinical effect, on both performance and health that easily go unnoticed. Once clinical symptoms occur, production has already been compromised. Moulds reduce palatability hence feed intake, which decreases production. They also reduce energy content (by 5–10%) as they utilise feed energy to grow. They can increase abortions and respiratory diseases in stock, and even respiratory problems in farmers when feeding out mouldy hay. There is no accurate way to estimate the amount of mould present, even using spore counts. The colour of the mould, say in silage, is a poor indicator of its degree of toxicity. Initially the problems may be negligible but within days or weeks the effects can become more pronounced. Mould can also lead to ketosis, displaced abomasum, oestrogenic effects (swollen vulvas and nipples, vaginal and rectal prolapse) which will reduce fertility. The adverse effects are amplified by-production stress, in that high-yielding cows are generally more susceptible than are low-yielding cows. Not all moulds are harmful or they may be in too low a concentration to cause problems. The rumen can degrade the mycotoxins to protect the stock. However the rate of degradation varies with type of mycotoxin. The extent of detoxification depends on rate of passage of feed, with dairy cattle having a rumen turnover eight times faster than beef cattle.

Urea toxicity

Urea contains 46% of non-protein nitrogen, which is the equivalent of 281% crude protein. As it has no other beneficial nutrients for ruminants, it will only be of benefit with rations deficient in protein but containing adequate energy. Urea is the cheapest form of 'protein' available but must be used carefully because of its toxic properties. It can be fed to stock in a number of ways but should never be relied on as the sole source of protein. Furthermore, it should only be fed to stock over six months of age.

Maximum daily intakes of urea should ideally not exceed 100 g/head/d for adult stock or 0.02% of live weight. At these high feeding rates, stock must be introduced to urea slowly over 7–10 days and fed in at least two batches per day. Grain based concentrates can contain up to 1.2% urea. Feeding very high levels, such 150 g/head/d, in once-daily grain supplements or PMRs, can lead to urea poisoning.

Urea supplements are often provided as blocks or as loose salt mixes and sometimes diluted with molasses in roller licker drums. Poisoning can occur if urea blocks have been softened by rain. Stock can develop a tolerance to urea if it is introduced gradually and they are given access to a dry mix of equal parts of phosphorus supplement and coarse salt for a week or so before urea is introduced; all ingredients are thoroughly mixed to prevent pockets of urea; and blocks or dry licks are covered or removed during wet weather.

Proprietary blocks are a convenient but expensive way to feed urea. Those containing 25–30% urea are of most benefit to cattle. Palatability problems may reduce consumption to less than 60 g/head/d. The major problem with these blocks is that there is no control over intake so not all stock will consume their requirement. It is much safer to feed urea incorporated into PMRs.

The symptoms of urea poisoning are rapid breathing, nervous excitement, lack of coordination, bloating and salivation. Because poisoning occurs quickly, treatment is often too late and ineffective. The remedy for urea poisoning is to drench affected stock with a mixture of 500 ml water, 500 ml vinegar and 1 kg sugar or molasses. Repeat the treatment if there is no improvement within 10 minutes. A relapse can occur several hours after the initial treatment, thus requiring further treatment.

Changes in milk composition

A drop in milk fat test tends to occur when the herd is placed on a low fibre diet (such as a diet high in cereal grain and very immature forage). When the rumen microbes ferment fibre, the resulting end-product, acetate, is used to produce milk fat. If the level of fibre in the diet is low, milk fat production decreases.

The easiest way to increase the fibre content of the diet is to directly feed out or incorporate hay or straw into the PMR. Take care though when feeding out poor quality forages because a drop in dietary energy intake could cause milk and protein yield to fall.

Low milk protein content is common in early lactation when cows are in negative energy balance. In other words, their energy needs are greater than their intakes causing them to lose body condition. Shortages of energy reduce protein utilisation by rumen microbes. As a result, the supply of microbial protein, cows' major protein source, is reduced.

Under most circumstances, providing a higher energy diet will lift protein test. Cow will only respond to protein supplementation with a lift in protein test if they are truly deficient in dietary protein. This is because they are unable to utilise energy properly when there is a protein shortfall.

When protein is lacking, microbial growth is depressed. As a result, microbial fermentation is reduced and less dietary energy becomes available. Cows then lose weight to compensate for the lack of dietary energy. When fat is mobilised, milk fat test tends to increase.

Monitoring manure consistency

Manure that is excessively loose or dry and firm for the diet fed may indicate a dietary imbalance that requires action. Manure pats can be easily evaluated using a 1 to 5 scoring system as described in Table 7.4.

When assessing the herd's average manure score, it is important to assess the range in consistency of faecal pats. If more than 20% are one score or greater (or less) than the average, this may indicate a nutritional imbalance or management problem. With cows restricted to sheds, the consistency of manure should only be made on faecal pats on clean floors, that is without any urine contamination that will change the consistency over time.

Table 7.4 Evaluating manure consistency.

Score	Manure description	Action required
1	Very liquid manure with consistency of pea soup. May leave cows rectum in a steady flow. Includes cows with diarrhoea	Increase effective fibre intake and seek nutritional advice
2	Runny manure which does not form a distinct plie. Manure will splatter on impact and form lose piles less than 25 mm high	Consider increasing effective fibre intake
3	Manure has porridge-like consistency. Forms a soft pile 40–50 mm high, which may have several concentric rings and a small depression in the middle. Make a plopping sound when it hits concrete floors and will stick to the toe of your shoes.	This is the desired consistency
4	Thick manure, sticks to shoes and readily forms piles more than 50 mm high	Consider reducing effective fibre intake and increasing concentrate intakes. Seek nutritional advice
5	Manure appears as firm faecal balls	Consider reducing effective fibre intake and ensure adequate drinking water is available. Seek nutritional advice

Metabolic disorders

Metabolic disorders can be clinical, when there are obvious symptoms, or sub-clinical, when there are not. Even at the sub-clinical level, they can depress feed intake and cause production losses.

Metabolic disorders such as ketosis and acidosis are usually linked to low intakes around calving or abrupt changes in diet.

Managing nutrition well during the dry period and in early lactation is the key to preventing or minimising the occurrence of metabolic disorders. The aim is to maximise nutrient intake around calving and in early lactation by providing enough high quality feed, and avoid decreases in intake caused by sudden changes in diet when cows calve and join the milking herd.

The most important metabolic disorders are summarised in Table 7.5 while one of the most prevalent disorders of intensive feeding management, lactic acidosis, is discussed in more detail later in this chapter.

Table 7.5 Metabolic disorders arising from unbalanced diets.

Disorder	Cause	Symptoms	Treatment
Milk fever	Sudden decrease in blood calcium levels.	Decreased intake and milk yield. Calving paralysis, death	Feeding management prior to calving to stimulate cow's ability to mobilise body calcium
Grass tetany	Low blood magnesium levels	Decreased intake and milk yield. Muscular staggers, death	Feed magnesium supplements
Ketosis or acetonaemia	Cows rely on fat reserves for energy during early lactation	Decreased intake and milk yield, Characteristic smell of breath	Feed well-balanced diet during early lactation
Lactic acidosis (grain poisoning) and laminitis	Rumen pH becomes very low due to high starch intake	Decreased intake and milk yield. See below	Include rumen buffers in diet and sufficient roughage
Bloat	Build up of foam in rumen which stops gas from escaping	Left side of cow is swollen. Cow stands up and lies down frequently	Put hose down oesophagus, administer oil, stab left flank to release gas
Urea toxicity	Ammonia poisoning	Rumen stops moving, death	Drench with water, vinegar and sugar
Feed toxicities	Anti-nutritional factors in diet. See below	Sickness and death	Identify cause and remove from diet

Maintaining normal rumen metabolism

Unless the rumen pH is maintained at a constant level, the rumen microflora cannot grow, multiply and function normally. Without normal microflora, digestion of dietary nutrients will be suboptimal, leading to depressed feed intakes and reduced cow performance.

When cows are fed high levels of cereal grains, starch is fermented quickly in the rumen to produce acids in the rumen. The production of lactic acid may be greater than the rate at which it can be absorbed or buffered. The resulting decrease in the pH of the rumen (increased acidity) stops other bacteria from digesting fibre. This in turn slows digestion and causes a loss of appetite. This condition is called lactic acidosis or grain poisoning.

Dietary fibre is an essential ingredient for this purpose as it promotes chewing, which stimulates saliva production thus recycling key rumen buffers to maintain normal rumen pH. When cows chew their cud (especially in response to fibrous forages) they produce large quantities of saliva (100–150 L of saliva/cow/day). With enough fibre in the diet, saliva production alone can generally maintain rumen pH.

Much of this fibre, however, has to be sufficiently long (greater than 1.5 cm) to stimulate this rumination or cud chewing. This is called long or effective fibre and ideally should make up at least 75% of the NDF in the ration. One good indication of the proportion of effective fibre in the diet is the proportion of cows ruminating when resting; ideally this should be at least 50% of herd.

This then is the logic behind formulating PMRs, so that each mouthful of feed contains virtually the same composition of ration ingredients hence concentrations of energy, protein and fibre. This minimises the diurnal peaks in rumen pH associated with pulse feeding of (often large quantities of) high-energy concentrates in the milking shed. Consequently, the digestion of feeds in the rumen and the resultant absorption of digested feed nutrients into the blood stream is relatively consistent throughout each 24 hr period. Further more, the stable rumen pH, allows for a stable population of rumen bacteria to optimise digestion of fibre and starch.

Acidosis can also be prevented in diets high in grain and low in roughage or fibre by supplying mineral buffers. Table 7.6 outlines some common buffers used in high grain diets. These are not a substitute for fibre, however, and the fibre content of the diet should be maintained.

Lactic acidosis

Acidosis can be clinical (with cows obviously sick) when rumen pH falls below 5.0 or sub-clinical when rumen pH falls below 5.5. Symptoms of sub clinical acidosis include low milk fat test, below 3.0–3.3%; low milk protein test; reduced milk yield;

reduced feed efficiency; sore feet due to laminitis or overgrown claws; manure in cows on same diet varying from firm to very liquid; manure foamy containing gas bubbles; manure containing larger than normal lengths of undigested fibre, more than 1.2 cm long; manure containing undigested yet ground grain; limited rumination, less than 50% of the cows cud chewing while resting; and cyclical feed intake.

To avoid acidosis, high starchy feeds should be introduced gradually (i.e. 0.5 kg grain/cow/day) so that the population of rumen microbes can adjust according to the type of fermentation that is required (more starch fermenting microbes may be needed). Different cows respond differently to grain feeding. Some cows can handle 6 kg grain/d while others will get sick on 3 kg/d; there is always a cow that will eat more than her fair share. The key to success is to make it a gradual daily increase and to watch your cows and check for symptoms of acidosis or grain poisoning.

Acidosis can be overcome by feeding more fibrous roughages, but that can lead to reduced feed intakes hence milk yields. Buffers can be included in the diet to stabilise rumen pH so that the rumen environment allows a healthy population of rumen microbes. Fibre and rumen buffers can easily be incorporated into PMRs.

Feeding management can also influence the incidence of sub-clinical acidosis in that when cows cannot eat when they are hungry, they overeat, having a larger than normal feed when they eventually get access to the feed trough. In this case the acidosis is not caused by lactic acid, but by excess production of the volatile fatty acids from rumen digestion. It is then important that all cows should be able to eat when they want to. Having feed continually on offer on a feedpad is then an important management consideration.

There is an extensive list (Dairy Australia 2007a) of feeding factors increasing the risk to acidosis and these include:

- *Type of cereal grain fed* – wheat is more rapidly digested in the rumen, barley or triticale, less rapidly while oats, maize and sorghum are slowly digested in the rumen.
- *Degree of grinding of cereal grain* – the finer the grinding, the more rapid the rumen fermentation rate, and the greater the risk of acidosis.
- *Amount of concentrate fed* – feeding concentrates at more than 5 kg/cow/feed or 10 kg/cow/d.
- *Presentation of the cereal grain* – incorporating the grain into a PMR, rather than feeding it while milking allows for more stable rumen pH
- *Total diet fibre levels* – NDF levels of less than 30%.
- *Type of grazed pasture* – young leafy pasture grazed in association with high supplement grain and low supplement fibre.

- *'Stale' wet supplements* – if using wet by-products (such as brewers grain, grape marc or waste vegetables) as fibre supplements, and they have been poorly stored, contaminated with mycotoxins or become less palatable, intakes will be depressed and rumen metabolism more prone to acidosis.
- *Ranges in length of dietary fibre* – with more than 50% of the fibre sources less than 1.5 cm in length, risks to acidosis are high. With greater than 75% of the fibre sources more than 1.5 cm long, these risks are low.
- *Insufficient rumen buffers in the ration formulation* – see the next section of this chapter.
- *Significant separation of feed ingredients in ration* – decreasing the opportunities of cows to consume sufficient fibre.
- *Daily feeding routine* – inconsistency in timing and proportion of daily ration fed out at any one time.
- *Restricted access to feeding area* – restricting access to feed or water can reduce intakes of each supplement.

Farms where cows are less able to lie down, hence spend too long standing, particularly on hard surfaces, can have greater problems with sore feet due to both trauma and acidosis. Cows should be able to lie down for at least eight hours each day. Other factors that can increase problems with sore feet include heat stress (when some cows prefer to stand); cows spending too long waiting to be machine-milked; and cows with 'perching' behaviour; namely, standing with their front feet in the feed trough and their back feet on the floor.

Mineral buffers and feed additives

Buffers stabilise rumen pH. The best example of a naturally occurring buffer is sodium bicarbonate supplied by saliva, in response to rumination or cud chewing. Mineral buffers are frequently incorporated into high grain diets to minimise the likelihood of lactic acidosis. Table 7.6 summarises those mineral buffers commonly used in milking cow diets.

Feed additives are another ration ingredient that has gained commercial attention in recent years. Their advertised benefits range from higher milk yields, to increases in milk fat and protein contents, to improved DM intake through a more stable rumen pH leading to improved fibre digestion. As with all feedpad diet ingredients, it is important to seek any relevant information to assist in making a value judgement that their inclusion is a good business decision. In other words, will the cost of their purchase and inclusion be lower or higher than any production benefits, under the existing farm and herd management conditions?

The primary feed additives registered for use in Australian dairy herds are ionophores and antibiotics. Monensin (for example, Rumensin®), Tylosin (Tylan®)

Table 7.6 Mineral buffers used in high grain diets.

Additive/ Buffer	% of diet DM	Feeding rate kg/t of grain	Function
Sodium bicarbonate (NaHCO$_3$)	1.5–2	15–20	Neutralises rumen acids to help prevent digestive upsets. Can be bitter and become unpalatable to stock if more than 4% fed. Tends to absorb moisture and form clumps which should be sieved out before feeding.
Magnesium oxide	Up to 1	10	Neutralises rumen acids. Source of magnesium to prevent grass tetany.
Sodium bentonite	Up to 4	Up to 40	Effectiveness as a buffer uncertain. Moderates the digestion of grains in the rumen and prevents cows from eating too much grain.
Crushed limestone	1.5	15	Effectiveness as a buffer uncertain. Useful in high grain diets as a source of calcium and magnesium.

and Virginiamycin (Eskalin®) produce their effects by modifying the rumen environment. These are all classified as rumen modifiers that alter the microbial population of the rumen, to change the mix of end-products from microbial fermentation. Other feed additives, such as yeast and yeast metabolites and Betaine, have recently been registered for use with dairy cows.

Monensin reduces the population of microbes that produce methane gas (which cannot be used by the cow as an energy source). The proportion of microbes that ferment feed to other more useful sources of energy is increased, resulting in improved milk yields. Responses to this additive depend on the diet and the stage of lactation. Doses for dairy cows are specified as 250–300 mg/cow/d.

Tylosin is an antibiotic which reduces liver abscesses (rather than prevent acidosis) and is recommended for use in combination with Monensin. Doses for dairy cows are specified as 150 mg/cow/d.

Virginiamycin is an antibiotic that inhibits the microbes that produce lactic acid and thus reduces the incidence of lactic acidosis. Doses for dairy cows are specified as 200 mg/cow/d. Being an antibiotic, it requires a veterinarian prescription for purchasing.

Yeast and yeast metabolites may assist fibre digestion and also allow cows to produce more glucose from the propionate (one of the volatile fatty acids) produced in the rumen. Betaine is currently being promoted to improve heat stress management by maintaining feed intake and reducing glucose lost in heat-stressed cattle.

8

Feeding management of partial mixed rations

This chapter discusses the sourcing and storing of feeds for partial mixed rations, the design of the feeding facilities and integrating the feedpad into the farm program.

The main points in this chapter:

- Efficient and effective sourcing of feeds requires a business approach of budgeting long-term requirements, then selecting the most appropriate feeds based on their nutrient costs, and finally, developing formal contracts with suppliers to ensure their continual supply at competitive prices.
- Feed wastage can be minimised by constructing the most suitable storage facilities, then managing them during storage and removal, and finally, designing the most efficient feed out systems. Managing the feedbunk properly is also a key factor in optimising cow performance.
- The dimensions of feeding face from silage stacks should be calculated prior to their construction, from the herd size and daily per cow feeding rate planning for a silage removal rate of 30 cm/day.
- Feeding and watering facilities should be designed to minimise competition between stock during the often restricted time they are on the feedpad.
- Feedpads provide the means of supplying supplements to grazing cows efficiently. The principles of effective grazing management should be followed to maximise the benefits of feedpad technology. Grazed pasture is the key to low cost dairy production systems in Australia.
- Computer software can assist with decision making in grazing and supplementary feeding of milking cows.
- To aid with planning and managing feeding practices, this chapter provides a checklist of the major considerations.

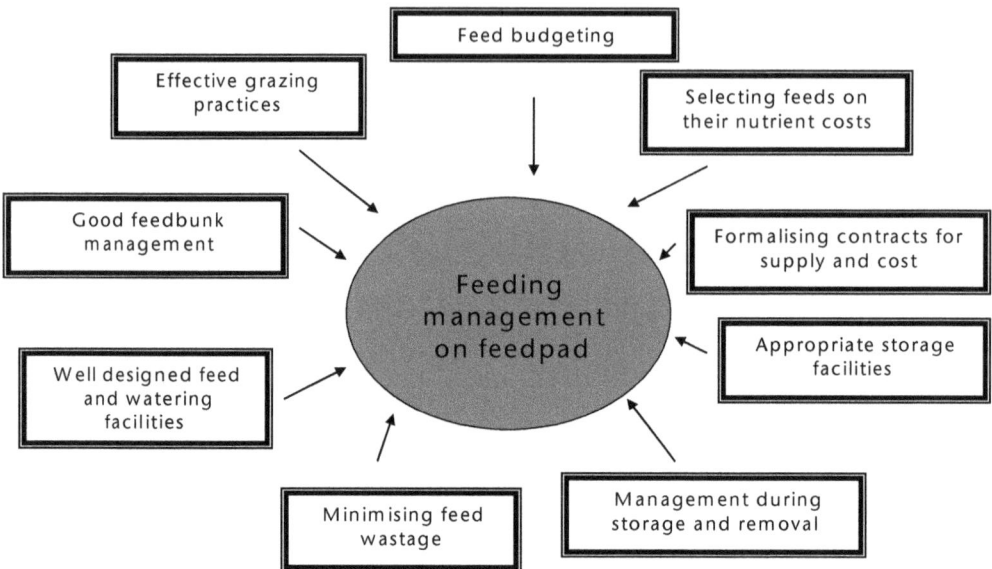

Figure 8.1 Some of the key considerations for good feeding management.

Appropriate feeding management is a key element for any well-run and profitable feedpad system (see Figure 8.1). It is one thing to provide well-balanced rations to supplement grazed pastures. These ration ingredients must be purchased at competitive prices then stored in such a way as to minimise post arrival deterioration, or for home conserved forages, post-harvest losses.

Feeding management comprises five main functions, namely purchasing the feeds or their ingredients, then their storage, handling, mixing and delivery. Storage can have a big impact on preserving feed quality and minimising losses, while the layout of the storage and mixing areas influences operating times hence labour inputs. Feedpad design also impacts on losses and labour but of most importance, it influences cow intake and performance.

This chapter reviews the mechanics of sourcing and storing feeds for partial mixed rations (PMR) and feeding them out to the stock. It also discusses the importance of design and management of feed and water troughs and integrating the supplementary feeding program with existing grazing practices.

Purchasing feeds

Preparing a feed budget

With feed costing 50–60% of the total costs for producing milk on most Australian dairy farms, feed budgeting should be an integral part of dairy farm planning. Month by month feed budgets should provide sufficient forward planning to take into account the seasonal variations in homegrown fodder production, the changing lactation cycles of milking cows and the feed demands of non-milking stock on the farm. A feed budget is simply a series of calculations of:

1. What are the total monthly feed requirements for the farm to achieve its milk production target, including those for dry cows and young stock;
2. How much of this feed can be grown on farm; and
3. Therefore how much of the monthly shortfall will have to either be purchased or taken from any farm fodder reserves.

It is usual to convert nutrient requirements, usually of ME, into pasture equivalents during step 1 and then reconvert them in step 3 to types of feed. This requires an estimate made each month of the daily ration composition of each class of dairy animal. Such estimates should take into account realistic allowances for feed wastages, based on the current feedpad system.

It is not easy to develop all-encompassing generic computer programs for planning feed audits for all dairy farms. One of these was developed specifically for irrigated farms in northern Victoria (KYFEED), and is available free from the author (john.moran@dpi.vic.gov.au or jbm95@hotmail.com).

Using the approach developed by Dairy Australia (2008a), the next step is to calculate the break-even price required for each of these feeds to provide a maximum affordable price that will still achieve annual financial goals. Such calculations should take in all other non-feed variable (operating) costs, as well as debt servicing, farm labour (including either paying the farmer and farm family for their imputed labour, or their monthly personal drawings) and other overhead costs. In addition, maintaining farm equity should be considered. The affordability of the purchased feed costs is linked closely to their ability to produce sufficient milk to cover the farm's overall financial plan. This exercise may require professional assistance but it will at least provide relevant break-even and target feed prices for purchasing feeds prior to shopping around.

Selecting the feeds

It is best to secure fodder supplies first, then calculate exactly what other feeds are required to balance the PMR. Formal contracts with suppliers of feed ingredients are the only reliable way of ensuring year round deliveries.

Depending on what it was purchased for, feed should be selected on the basis of the cost per unit metabolism energy (ME), crude protein (CP) or neutral detergent fibre (NDF). Initially all feeds must be compared on a dry matter (DM) basis. The relevant calculations are as follows:

$$\text{DM comparison: c/kg DM} = \frac{\text{\$/t fresh} \times 10}{\text{DM\%}}$$

$$\text{Energy comparison: c/MJ of ME} = \frac{\text{c/kg DM}}{\text{ME}}$$

$$\text{Protein comparison: \$/kg of CP} = \frac{\text{\$/kg DM}}{\text{CP\%}}$$

$$\text{Fibre comparison: \$/kg NDF} = \frac{\text{\$/kg DM}}{\text{NDF\%}}$$

These calculations are included in a computer program (MIFC10) available from the author (john.moran@dpi.vic.gov.au or jbm95@hotmail.com).

From knowledge of the actual price of each feed nutrient, a more objective weighting can be placed on its value for milk production. For example, if the barley grain in Table 7.2 (with 88% DM and 12 MJ/kg DM of ME) could be purchased for $250/t fresh, its energy would then cost 2.3 c/MJ. If farmers were looking for another energy supplement, they would need a convincing reason to pay more than this. It also means that every 1 MJ difference in ME is worth $23/tonne. Therefore when considering other feeds, such as a hay with feed analysis data available, then one with an ME of 10 MJ/kg DM is worth $23/t more than one with only 9 MJ of ME/kg DM. This is then what farmers should offer to pay extra, provided feed analysis data were available for both hays.

Likewise to obtain a benchmark of the cost of protein, if the cottonseed meal in Table 7.2 (with 90% DM and 43% CP) could be purchased for $450/t fresh, its protein then would cost $1.16/kg. Therefore a 1% difference in CP is worth $11.60/t or a legume hay with 19% CP is worth $11.60 more than legume hay with only 18% CP.

Having a dollar benchmark figure like this for every 1 MJ difference in energy or every 1% difference in protein is convenient when comparing the nutritive values of a range of energy rich or protein rich supplements.

Other considerations when sourcing feeds include there must be a good reason for selecting each feed, whether for energy, protein of fibre, to achieve herd production targets; when selecting some feeds, such as fresh fodder, the costs of shrinkage during storage or wastage during feed out should be taken into account; if choosing feeds to supply NDF, ensure sufficient fibre is long enough (greater than 1.5 cm) to effectively stimulate rumination? Also, ask the following questions, can the current feed storage and mixing facilities handle this new feed? And in addition to the nutrient analysis, what else is known about the feed? If it is a conserved forage, how was it managed during growth? Was it likely to have been drought-stressed, which could influence plant nitrate levels. For all purchased feeds, particularly agro-industrial by-products, a Vendor Declaration Form should be sought to ensure it contains no chemical residues. How has it being stored prior to sale? What is the likelihood of spoilage increasing feed nutrient costs and maybe adversely affecting palatability?

A new tool has been developed for farmers to use of their feed analysis in making better feeding decisions on their farm. The 'MyGrains2Milk Feed Report' provides star ratings to put these feed analysis into context of what is high and low quality for that particular feed. It also converts feed prices into c/kg DM, c/MJ of ME and $/kg CP, based on the laboratory report and feed's purchase price. In addition it provides practical tips on buying, storing and feed out. The tool is available for free at www.myG2Mfeedreport.com.au. Try and view the feed prior to purchase to ensure its integrity.

There are many things to look for when selecting feeds such as:

- *Cereal grains* – excessive small grains, weather damage or mould, poor uniformity of grain mixes.
- *Pelleted feeds* – excessive dust, loads delivered still warm.
- *Hays* – visual signs and odours indicating it was baled too wet, poor leaf content.
- *Silages* – DM content too high or low for good fermentation, excessive chop length, bad odour, mould.
- *By-products* – inconsistency in appearance between deliveries, DM content too high or too low, contaminants or foreign materials, mould.

Purchasing the feeds

The next step of the feed buying process is focusing on the three key issues of feed quality, supply and price (Dairy Australia 2008a). Prices for cereal grains in Australia are dependent on international supplies and demands, hence global prices for feed grains. These also have to take into account demands for food (for human consumption) and for fuels (ethanol and biofuels). Seasonal climatic conditions can also have a big influence on grain prices making predictions very inaccurate. There are independent grain brokers that can provide the best market intelligence for medium to long-term planning.

Unlike grain, Australia's fodder market is domestic, and is less transparent because most conserved fodder is traded directly from farmer to farmer, rather than through formal brokers. Over 70% of dairy farmers purchase their hay and silage direct from another farmer, although over 40% also use fodder merchants or traders. Furthermore, in Australia, over 60% of the hay, silage and straw is traded within the dairy sector. Therefore good relations with fodder suppliers and traders are critical for dairy farmers to manage their quality, supply and price risks. Written contracts provide more security than loose 'word of mouth' agreements. Formal agreements vary in type from immediate spot, to short-term forward, to longer term rise and fall contracts. There are also differences in pricing strategies. Forward purchasing may seem more expensive but it does transfer price risk to the seller while the seller also retains the production risk because he is obliged to deliver the feed.

The majority of farmers rely entirely on spot purchasing, buying their feed month to month as required. Some dairy farmers however, manage their exposure to grain price movements using short and long-term forward contracts and other methods such as hedging using the grain futures market, as more sophisticated purchasers like the stockfeed companies and intensive livestock producers do.

It is important to maintain a buffer against poorer forage growing seasons. In recent years, more farmers are prepared to carry these extra costs (for storage and unused farm resources) for such on hand forage resources, with pit silage being the preferred long-term stored fodder.

The 2008/9 global economic crisis placed downward pressures on export milk prices (see Figure 2.4), while the series of drought years in southern Australia has placed upward pressures on hay and grain prices. Consequently there is even greater need for farmers to ensure they gain maximum milk responses from money invested in feed purchases. This can best be achieved by including specifications of hay or grain quality in any formal agreement and rejecting any loads that fall below the specified feed quality. In addition they must check 'the small print' of any signed contracts carefully for the possibilities of renegotiating purchase prices and delivery dates, if the need arises.

Grain and fodder can be purchased directly (and frequently more cheaply) from a producer rather than a feed trader or merchant. The supply chain costs, market volatility and supplier risk may negate any benefits. The logistics of and costs for feed storage until required on the dairy farm must be taken into account.

Undertaking a monthly feed budget provides farmers with negotiating power to lock in a supply schedule for delivery over several months, with the current feed prices applying for each month on delivery. This effectively makes time of supply and time of pricing independent of each other. In addition, terms of payment can be negotiated with feed suppliers to suit cash flow and milk cheques.

Reducing feed wastages

The costs of feed wastage are two-fold, firstly the cost of feed purchased but not consumed by the stock, and secondly, the reduced feed quality reducing the milk produced per kg feed actually consumed. The economics of feed wastage are discussed in Chapter 11. Feed wastage is probably one of the hardest aspects of feedpad management to quantify due to lack of accurate information from data actually physically collected, particularly from dirt and gravel feedpads. Without any photographic guidelines at present, this makes it even more difficult to visually assess to reduce the degree of subjectivity when incorporating such data in economic appraisals of various feedpad systems.

Offered feed may not all be consumed by the stock for a variety of reasons. It may not be palatable when fed out so is voluntarily rejected. It may become unpalatable once out on the feedpad, for example due to rain. Feed might become contaminated by soil, manure or other contaminants. This is usually called trampling losses. Feed out chutes and speed of delivery from mixer wagons are not calibrated to the physical dimensions of the feeding area. The frequency of feeding out is not proportional to the time stock spend on the feedpad. The stock cannot physically reach the feed, or the amount of offer greatly exceeds the stock's appetite. Trampling losses can include fine material which has filtered down to the base of the feedpad as well as courser material which becomes contaminated because is has been spread around by stock in the process of consumption. Baled hay or silage is likely to have greater wastages than chopped conserved forages, because more material is likely to be dropped and remain uneaten, especially in wet weather.

Feed losses occur during delivery and storage, mixing and feed out. As feed out losses are significant and costly, investments in improving feed delivery are the most worthwhile. Even within a feedpad system, feed wastages can be reduced by preventing stock walking over and defecating and urinating on the feed; improved maintenance of feeding out equipment; appropriate trough design and management; timing of feeds within any 24-hour period, so excess feed is limited; the ability for cows to sort through the feed; feed quality and palatability; effects of climate; and cow appetite

All feeds should be stored to minimise spoilage and wastage. Inexpensive feed storage facilities may work in the short term but they usually lead to higher feed spoilage and wastage costs. Concrete surfaces and commodity sheds are essential for long-term storage of expensive PMR ingredients. Other factors to take into account to reduce storage losses include insurance; insure anything you cannot afford to lose and that means hay, both homegrown and purchased. Laboratory analysis of feed quality is the only way to be sure you get what you pay for. These values can change with different loads. Ensure there is easy access to feed storage areas, with no power lines likely to cause problems with large trucks. Dry ingredients must be kept dry and protected from gusts of wind. Wet ingredients can shrink by up to 25% after delivery due to seepage and biological activity while in storage. Wet protein sources can be subject to fly infestation. Contamination with stones or dirt can be a problem in storage areas without a concrete base. Some feeds contain anti-nutritional factors (ingredients which reduce palatability or efficiency of utilisation) and these can include toxicities, as was discussed in Chapter 7.

Storing the feeds

Feeds can be stored in a variety of ways, such as above or below-ground bunkers with removable covers; steel or concrete vertical silos; inside sheds; in tanks, for liquid feeds; and in bales, for hay and silage (see Figures 7.2 and 8.2).

Feeds should be stored in secure, vermin-proof and well-ventilated locations with easy access for feeding out machinery and cleaning. They should be located at least 20 m from the dairy and well away from waterways. Ideally any feeds that can generate offensive odours should be frequently sourced for minimal storage times. Silage liquor is strongly corrosive so any leachate or runoff should be directed to the effluent system.

Conveying and distributing feed can lead to spillages adding to odours, contaminating runoff and attracting vermin. Stormwater from such areas should be diverted to the effluent system.

Feeding out equipment should be operated to minimise spillage. Good feeding management implies regular cleaning of feed troughs and other feeding areas to ensure stock are only offered freshly mixed feed. When feeding out on flat surfaces, cows will often push the feed out of their reach, so a food scraper should be used to push it back to the cows about 30 min after the cows start feeding. Rejected feed can be fed to dry stock or even composted for storage.

It may be worth considering having fences within the feedpad to split herds into smaller groups for feeding, such as first calf heifers and older cows.

Regular maintenance of the feedpad is essential. Feed residues should be removed from the feeding area, while areas under fences and around the feedpad should also be cleaned, at least weekly. Water troughs and float valves should be checked monthly to minimise overflows. The feedpad surface, drainage channels and effluent storage should also be routinely maintained.

Commodity bunkers

These should be located close by the feedpad with good road access for large trucks. They can be open or under cover. Having a series of open bunkers adjacent to each other reduces the number of walls to construct. They should have concrete floors and brick, concrete block or prefabricated or reinforced cement walls. For storing by-products with very low dry matter content, soil ramps adjacent to these storage bunkers will aid discharge from trucks. Typical bunkers to store semi-trailer loads of commodities are 6–10 m in depth, 4–6 m wide, with walls that are 1.5–2.5 m high and 150–200 mm thick and with a 5–7 m concrete apron in front of each bunker. A railway iron post can be installed at the beginning of each bunker wall to protect it from vehicle damage. Roofs should be of sufficient height (4–6 m) to allow sufficient working space for front-end loaders and maybe unloading of semi-trailers, and with sufficient eaves to provide protection from rainfall. The roofs could even be constructed to slide in and out to aid vehicle access yet reduce spoilage during storage. Floor slopes may be back to front for drainage of rainfall, or if under cover, they may even be from front to back when designed to store very low dry matter material.

Liquid commodities such as molasses should be stored in tanks designed for ease of filling and emptying into mixer wagons. Hay can be stored outdoors provided it can be covered for protection against rain.

Considerations when planning commodity bunkers include:

- Farm
  - location and accessibility to feedpad
  - machinery and equipment requirements
  - minimising wastage.
- Environment
  - supplement leakage
  - odour
  - farm waste, silage wrap, bale twine.

Storing silages

Following efficient harvesting of high quality crops for silage, there can still be significant losses of DM and quality if the storage system is inadequate. These losses can be due to excessive respiration (overheating), effluent loss and aerobic spoilage in the stack or bales. They can be minimised by good management during filling and storage. There is a range of storage systems used for preserving silage which include underground or above ground systems, with the capacity for handling both chopped and baled forage.

Forage harvested silage can be stored in above-ground systems, such as bun stacks, moveable clamps, bunkers (Figure 8.2) and clamps, tower silos, vacuum storage or stretchable plastic bag; or in-ground systems such as trenches or pits.

Baled silage can be stored in groups, in rows, stacks, stretch wrapped in line (see Figure 6.7), stretchable plastic bags or in a pit or trench, or individually; that is, wrapped individually or in double bale bags.

Figure 8.2 Cement bunkers minimise silage wastage.

Above-ground systems rely on a plastic cover for protection but these generally only provide short-term storage. Storage time can be increased to several years by providing a second protective cover over the silage plastic, to reduce breakdown by ultraviolet radiation from sunlight. The system chosen will depend on the purpose for which the silage is being used, available equipment, method of feeding out and personal preference.

Buns or stacks are silage units formed above ground, with no structure support. They have no construction costs, are very adaptable for self-feeding using electric fences and can be easily sealed using a front-end loader blade or bucket. As they have a high surface to volume ratio, plastic costs per tonne silage and wastage rates can be high.

Above-ground bunker or clamp silos are the most common storage systems and their cost depends on the material used (earth versus concrete floor, earth versus post and weldmesh versus concrete walls). They can be built to open at either end and even be adapted for self-feeding. In bunkers with portable clamps, the walls can be removed if so desired, once the stack is completed. Using large hay bales as alternative side walls is not recommended as it is difficult to obtain an airtight seal. Tower silos are very expensive and rarely used in Australia.

In-ground pits are the best for long-term storage as they can easily be covered with soil. They cannot be made in areas with high water tables nor can they be used during prolonged wet weather. Other forms of in-ground storage include hillside pits (which are constructed into the sides or tops of hills) and trench silos (where silage is stored partly above and partly below ground level). These provide for good compaction and can be used for self-feeding if the base is solid and dry and can be scraped clean.

For baled silage, stretch wrapping and stretchable bags provide flexibility in storage siting and a relatively small exposed feeding face. But they require contractors, are expensive forms of silage storage and rodents can be a problem.

The volume of silage storage needs to be determined once the quantity of greenchop produced on farm or to be purchased is decided. Silage density depends on many factors, such as DM content, chop length, type of silage storage system and the degree of compaction in the stack.

Another key factor when storing silages is the liquid produced during storage, called leachate. Silage leachate is a highly toxic, 40 times stronger than dairy wastewater, and lethal to aquatic life. The amount produced on the moisture content of the greenchop being ensiled. For unwilted grass, it is over 100 L/t while for grass wilted to 25% DM, it is less than 30 L/t. Ideally, it must be contained and managed so it does not enter ground or surface waterways. When managing silage leachates, consider that silage must be stored at least 20 m from waterways, 45 m from the dairy and 100 m from neighbouring houses and groundwater bores; it should be sealed to prevent seepage into the ground; the leachate should be contained with at least 3 m^3 storage per 100 t greenchop; it should be mixed with dairy effluent before being disposed onto land. Also consider that if the leachate is fresh and originates from good quality silage, it can even be fed to stock. Design silage bunkers with sloping floors that link up with a drainage channel across the front of each bunker. Epoxy resin can be applied to any concrete feeding area to prevent corrosion.

Detailed designs of storage systems and their management to minimise silage losses during feeding out are beyond the scope of this book. These are discussed in detail by Moran (1996) and Kaiser *et al.* (2004).

Calculating the dimensions of the silage feeding face
It is important to plan the physical dimensions of silage bunkers or stacks carefully prior to their construction. This should be based on the rate of daily removal of the silage. Silage density depends on many factors, although typical densities for silages with 25% or 35% DM are 680 and 630 kg/m^3 respectively.

The minimum silage removal rate across the stack should be at least 15 cm/d, with 30 cm/d recommended with unstable silages and in hot weather. The surface area of the feeding face required to achieve target feed out rates can then be calculated from the quantity of silage fed out each day and the silage density. The average amount of silage fed out per day is determined by the size of the herd and its daily silage requirement. The width of the silage storage depends on it height. Such calculations should be undertaken prior to constructing the silage storages. An example of the mathematics required is as follows:

$$\text{Stack width (m)} = \frac{\text{Daily silage removal (kg fresh/d)}}{\text{Silage density (kg/m}^3\text{)} \times \text{Stack height (m)} \times \text{silage removal rate (m/d)}}$$

A herd of 250 cows will be fed on average 5 kg DM/day of pasture silage (with 30% DM content) to be stored in a bunker 2.5 m high with a silage density of 650 kg/m^3. The silage removal rate should be 30 cm/d.

$$\text{Stack width (m)} = \frac{4170}{650 \times 2.5 \times 0.3}$$
$$= 8.5$$

The herd's requirements are then 4.17 t fresh silage/d and the stack should be 8.5 m wide.

Storing hays

A well-built hay shed is the best method of protecting stored hay from the weather for up to three years. Losses are minimal when bales are well stacked. The shed need not be fully enclosed. So long as the sides facing the direction of the prevailing wind and rain are covered, the outside bales are protected from weather damage. Each tonne of baled hay needs 6–7 m^3 of shed capacity.

When constructing a new shed it is important to consider the distance from the feed out area and the ease of access of hay-handling machinery. The bottom bales need to be insulated from contact with soil moisture using a concrete floor, gravel covered by a plastic sheet, wooden slats or even old corrugated iron sheeting.

Hay stored in the open can lose as much as 20% of its DM if stored over the following winter compared to only 6% if used beforehand. These losses can increase to 40 or 50% in high rainfall areas. DM losses in 1.8 m round bales stored outside are only half those of the smaller 1.2 m round bales.

When storing hay in the open, the site should be level, well drained and safe from fire. The narrow end of the stack should face the direction of the most frequent rain. Stacks should always be built to a set plan so that bales in each layer can bond together. When building a stack by hand or using mechanical equipment, the bales should be pressed firmly together with alternate layers at right angles to those below.

Small stacks with bales in the foundation layer laid on their flat side can go as high as 10 or 12 layers. If the stack is too high, the additional pressure from the extra bales can cause the ageing outside bales in the foundation layer to collapse and the stack to fall apart. With larger stacks, the foundation bales should be laid on their edges with cut sides down, making them more rigid and less likely to crumple under the pressure of bales stacked above them. The shape selected for the roof will determine the way the bales in the roof are stacked. The roof can be made from galvanised iron or plastic sheeting tied down securely with ropes attached to stakes at the base of the stack or bale twine on bales in the foundation layer.

Moist bales should not initially be stacked tightly together but small gaps should be left between them to aid air circulation. Large bales should not be forced into small centre gaps as they might dislodge the edges. It is easier to fill these gaps with broken bales. The sides of the stack should be kept perpendicular or inclined inwards. The steps in the roof bales could even be filled in with straw or broken bales to give a good foundation for the roof.

When stacking large rectangular bales, their sheer size gives them a natural stability. Three layers can be stacked on top of one another without the need for interlocking designs.

Whether storing hay in shed or in the open, it is worthwhile recording from where particular groups of bales originated. This, together with other information about the particular pasture or crop cut for hay, its management prior to harvest, and any other details about the conservation procedures, such as the weather during field curing, may be useful when assessing the quality of the hay and/or the animal performance when fed out.

When storing round bales in sheds, it is advisable to wait for several days after baling before stacking if concerned about the bales heating up. If space is available, the bales could be spread out in a single layer and then stacked after some drying has occurred. They can be stacked either on their ends or their sides. Bales stacked on their ends provide a more rigid and less compressible framework. Stack height depends on the ability of the foundation bales to withstand the weight of bales above them and limitations of the mechanical stacking equipment.

Round bales are usually stored in the open. Drainage is an important consideration in high rainfall areas, so it is best to use a well-drained base such as crushed rock, old tires, wooden pallets or old poles. Ideally bales should be lined up in single north–south rows so their sides have maximum exposure to sunlight. The most common method is to make rows of bales, the bales of each row being lined up in a long cylinder with their ends touching. Bales in different rows should not touch each other because any water accumulating in the valley between them might encourage mould growth. A minimum of 60 cm should be left between rows, not only to allow the water to drain off the bales and to drain quickly, but also to ensure adequate ventilation. If stored in pyramid stacks, water from the upper bales will flow into the lower bales.

Bales should not be stored across the slope of hills as they will stop and absorb water runoff. They should not be stored under trees since trees tend to reduce the drying effect of the sun and wind and may also drip water onto the bales long after the rain has stopped. There should be a substantial fire break around the storage area. Bales should never be stacked close to road intersections where they can block the view of motorists.

Large plastic sheeting will help reduce weathering losses, but if planning to cover round bales, it is best to stack them aligned in rows end to end. The sheets should be held in place using ropes fixed to anchor points on the ground or to a strained wire wrapped around the entire stack. Plastic sheets over single rows or pyramid stacks of bales should be placed in such a way that rain falls onto the outer extremities of the bales or straight onto the ground.

With round bales stored outside, practically all the losses of feed quality occur in the external weathered layer, or thatch, that develops when the bale is first weathered. This layer also helps to shed rain and protect the inner portion of the bale from damage. Its resistance to weather varies with the degree of coarseness of the material in the bale; the coarser the material, the deeper the thatch cover. As the thatch layer ages, its surface gradually decomposes, allowing water to penetrate deeper into the useful hay in the bale. This process is accelerated in high rainfall areas. For round bales 1.5 m in diameter and 1.5 m wide, a thatch of 8 cm makes up 21% of the round bale.

The portion of the round bale in contact with the ground, the under roll, is very vulnerable to damage from microbes causing it to decompose. In wet environments, six months is probably the optimum storage life for unprotected round bales in the field. The rate of deterioration of hay in such bales is up to 20% per year compared to less than 10% in round bales stored under cover.

Hay fires

If hay has a DM content of less than 75%, it can heat in storage, causing a loss of nutrients. Once the temperature exceeds 45°C, it can cause a fire capable of destroying both the hay and the hayshed. The danger arises when pockets of air trapped in the wet hay allow the respiration of the fungi to increase temperatures to 45°C. This heating continues even further if the fungi are replaced by heat-resistant bacteria when the temperature rises to 70°C. If the hay remains wet, and the generated heat remains trapped, further chemical reactions can boost the temperature to 200°C at which point the hay begins to smoulder. Outside, the moist hay surrounding the hot spot will cook and bond with the burning hay, forming a cake that seals in heat and gases. Eventually the pressure of the gases breaks open the stack and the smouldering hay makes contact with oxygen from the outside air and the hay bursts into flames.

Hay can be made wet if rain enters the stack or if ground water damages the base of the stack. All freshly built stacks should be checked in the first two or three days after storing and then at gradually lengthening intervals for the next eight weeks, the main danger period.

The stack temperature can be checked by driving a piece of 20 mm water pipe into the stack at an inclined angle where heating is suspected and measuring the temperature by lowering a thermometer down the pipe. For safety sake, the thermometer should not read above 70°C. Alternatively, a sharpened steel rod can be driven into the suspected hot spot and left there for several hours. After if is removed, feel for the hottest spot along its length. At 45–55°C, the bar is tolerably hot, at 55–60°C, the bar can barely be touched while above 70°C, the bar is too hot to touch. Other signs of overheating are condensation and an acrid smell.

If the temperature continues to rise above 70°C and cannot be controlled, the only solution is to make sure adequate fire-fighting equipment is on site then dismantle the stack. Never walk on hay that is likely to be smouldering because it could collapse under your weight.

Heating can be minimised by reducing the compression of the bale, provided the hay is not intended for sale. The bale length may also have to be decreased to reduce handling problems of loosely baled hay that more easily falls apart. Bacterial additives can reduce heating, particularly with lucerne hay. Small moist bales can be placed in little stacks in well-ventilated storage areas, leaving plenty of space between bales. The circulating air absorbs moisture and heat providing a cooling effect.

Large and small rectangular bales react to heating in much the same way but round bales react differently. They may only reach slightly higher internal temperatures at the same DM content, but because of their size, the heat is retained longer in large round bales. They then should be baled at higher DM contents than rectangular bales. The coarser the material in the bale, the easier it is for generated heat to escape.

If mould growth heats the hay to 40°C or higher, severe browning reactions begin. These reactions combine amino acids and sugars to reduce the availability of the N for rumen digestion. The acid detergent insoluble N (or ADIN) level gives a guide to this unavailable N. These reactions also increase acid detergent fibre contents of the hay thus reducing digestibility

With hay harvested and stored at less than 80% DM, DM losses and quality deterioration are much greater than in drier material. Very moist hay stored at 64–75% DM can lose 14% of its digestible DM during six month storage. Heating of hay will also increase when bale density is higher.

During storage, hay often loses further moisture until its DM content stabilises at 80–95% DM, depending on temperature and humidity. For example, hay stored at 30°C will eventually reach 93% DM in dry regions with relative humidity averaging 30%, whereas it stabilises at 82% DM in humid regions where relative humidity averages 70%. If hay is sold on a bale-weight basis, this loss in moisture is loss of income.

Storage losses are caused by continued plant respiration, by activity of microbes and by chemical oxidation of plant material. The magnitude of such losses is influenced by many factors such as hay DM content, storage facility or site (hayshed, open stack or covered with plastic), bale size and density and length of storage.

During the 2007 and 2008 spring hay-making seasons, a lot of cereal hay was made from failed cereal grain and canola crops in southern Australia. There was also a large increase in hay fires from one to four months after baling with most of the fires in large square, rather than round bales. These were attributed to higher than normal plant sugar levels, the difficulty of curing the thick stems of canola crops, the results of an urgent demand for hay, hence rush to bale the crops before proper curing. As the crops failed to produce grain because of extremely dry seasons, they had above-normal levels of plant sugar levels in their stems. This could have led to higher then normal plant respiration and greater microbial activity during storage, leading to excessive heat generation within the stacks.

The likelihood of hay fires can be reduced through application of hay preservatives. These reduce microbial activity and mould growth in high moisture hays thus allowing baling at higher moisture contents. This results in shorter curing times, reduced leaf shatter and leaf loss (thus improving hay nutritive value). They also help maintain the hay's green colour and palatability for stock.

Feeding out conserved forages

Feed out systems can range from the very basic self-feeding silage from a pit or feeding whole bales of hay or silage in the paddock through to the expensive integrated feedpads described in previous chapters.

Removing the forage from storage

Removing hay from stacks or sheds is quite straightforward because the conserved forage should not further deteriorate. The anaerobic storage of sealed silage, however, ends once it is exposed to the air. Silage is a very perishable product requiring rapid transport to the stock. The rate of spoilage depends on many factors such as the speed of its removal from storage, the equipment used, the skills of the operator and its accessibility when presented to the stock.

When planning storage and feed out facilities for silage, it is important to identify the number of stock to be fed, the likely period of feeding and the quantities to be fed out. Efficient high-throughput systems are necessary when handling large quantities of conserved forages, whereas basic facilities will suffice for feeding out small quantities. As well as capital investment, acceptable feed out losses and labour use efficiency (hr labour to feed out each tonne of forage) should be considered.

Removing silage from storage requires serious attention because the management of the feeding face can have long-term impacts on the silage's aerobic stability and wastage. The first sign of spoilage is the heating of the silage. If the silage is unstable, aerobic losses can exceed 30% DM and lead as well to large losses in nutritive value (due primarily to energy lost as heat and heat damage to proteins), with poor palatability reducing voluntary intake, thus increasing wastage even further. Nutrient losses are invariably greater than those of DM, up to 40–70% higher for silage energy (Kaiser *et al.* 2004). Excessive aerobic spoilage can be addressed by good management during silage making, such as rapid filling, good compaction and effective sealing of bunkers; using silage additives to improve silage stability once the stack is opened; ensuring good management during feed out, such as having a high rate of silage removal and minimum disturbance of the feeding face to prevent air penetration. Silage removal rate is usually quantified as the speed at which the feeding face moves into the stack over time, in cm/d. Remember to minimise disturbances of remaining silage in the stack. This can be best achieved using a silage block cover or a tractor mounted shear grab, rather than a front-end loader, but such equipment is expensive and only single purpose. Also, ensure the plastic covering the stack is only pulled back as far as the daily silage is removed. It may also be worth considering pulling the plastic back over the exposed face.

Feeding the forage to stock

Baled hay or silage is usually removed from the storage site using forks or a spike mounted on a front-end loader or to a three-point linkage. Trailers will increase the number of bales, hence the work rate when the storage and the stock or mixing facilities are some distance apart.

Chopped silage requires mechanical removal from the stack and if required, transportation to a mixer or silage feed out wagon. Once out of storage, baled and

Figure 8.3 Metal 'skirts' at the bottom of hay ring feeders allow for the feeding of chopped silages.

chopped hay and silage can be presented in many ways, such as self-feeding from a hay shed or the silage face, using an electric fence of feed barrier to restrict access; self-fed from a flat-top trailers; windrowed on the ground in a paddock; bales can be fed out whole in the paddock; bales can be fed whole in a feeder or using a feed barrier, or fed via feed trough or conveyer belt through an electric fence, or onto a feed pad.

Design of water and feed troughs and feeding strips

Feeding and watering facilities should be designed to minimise competition between stock to allow maximum intakes during the often restricted time they are on the feedpad.

Water supplies

Daily water requirements are discussed in Chapter 6. Stock must be provided with sufficient drinking water at all times and the system should be able to supply at least 20 L/cow/hr to meet likely peak demand. The optimum temperature for drinking water is 15–17°C. Peak demand often coincides with feeding times. The pipe diameter needs to be at least 75 mm with an operating head of at least 10 m. A tank could be used for short-term supply in the event of a power failure. A back-up water supply should be available to hold at least two days' peak requirements, in case of a breakdown or loss of normal water supply.

Upgrading the facilities can involve increasing the rate at which troughs fill or providing greater trough capacity. Options involve upgrading pumps and water pipe diameters or increasing the number of watering points, particularly those close to the feeding area. Cows often drink a third of their daily water intake in the hour after morning milking on a hot day.

When depending on dams, water quality can become an issue because as dam levels drop, salt contents, bacterial loads and algal outbreaks all tend to increase. Grazing cows can tolerate reasonably high levels of salt in their drinking water, but tolerance levels drop as they are offered less pasture. Problems that limit water intake also limit feed intake, hence cow performance.

Water troughs should be well separated (but within 15 m) from feed troughs with the same water trough servicing adjacent pens. Provision should be made for water spillages, or leakages, to directly enter the feedpad drainage system and also for drainage control in the event of burst mains or a jammed float valve. Troughs should be located at the high point of a water line to reduce sediment and to facilitate purging air from the pipeline. Trough design should allow for regular, easy cleaning with a removable bung for complete drainage. In free stall sheds, they should be located at crossovers to reduce the incidence of stock blocking each other in alleys.

Each cow should be provided with 75 mm of linear watering space in free stall sheds while for circular water tanks, one watering space (60 cm of tank perimeter) should be available for every 15–20 cows. A water depth of 15–20 cm helps keep water cooler, fresh and easier to clean because less debris accumulates. The optimal trough height is 60–90 cm, from ground to the top of the trough. Watering points should be cleaned out at least weekly to remove any feed and other contaminants.

Access to a reliable supply of water of acceptable drinking quality is then imperative. Watering systems may also be required for dust suppression and for sprinklers to cool cows. Although such runoff may be minimal, it should be directed into the feedpad effluent system. To source water from waterways or groundwater, official licences may be required.

Feed troughs and feeding strips

It is important that all stock can eat comfortably with minimal competition. Whether feeding into troughs or onto cement feeding strips, 70 cm feeding space should be allocated per mature cow. For a feeding strip or trough when cows eat from both sides of the strip hence face each other to eat, this equates to a total feeding strip allocation of 35 cm/cow. Trough space can be reduced to 45 cm/head for 6-month-old cattle, or to 55–60 cm/head for stock 18-months old. Competition between animals is much less severe when complete rations are fed *ad lib* throughout the day, so trough lengths can be reduced, even down to 30 cm/head.

Ideally, troughs should be covered with a grate that directs the cow's head downwards and precludes the lifting of the head above the trough; however, this feature must be considered alongside the ease of filling the trough with feed. Lockable feeding head bales will reduce the incidence of cows throwing feed around as they lift their head, hence reducing wastage. In the latest feedpad designs in the US, concrete feeding strips are covered with epoxy resin to reduce corrosion of concrete. Feeding strips should be positioned 75–150 mm above cow feet level with a nib wall 400–550 mm above the feeding strip, to protect the feed from the effluent. Trough width should be related to the reach of the animals which is up to 80 cm for mature dairy cows eating from the base of trough 30 cm above ground level. For ring hay feeders, a 2.2 m diameter ring (Figure 8.3) will hold up to 300 kg of hay or 800 kg of silage and will cater for 18 cows.

Feed troughs and strips should be situated on the high side of the pen, running parallel with the contour to minimise pad drainage. They should also have smooth surfaces, as those without grooves or holes that can trap feed are easier to clean and help reduce build-up of waste feed, mould growth and unpleasant odours. The base of the trough should then be raised 10–30 cm, with the front 50 cm above this. If too much room is allowed, wastage can increase as the objective is to allow the stock to feed and then move back. Feed barrier walls wider than 15 cm can restrict how far stock can reach into troughs or onto feed strips.

The height of any wall constructed to retain feed should accommodate the chute of the feeder wagon which should be higher than 60 cm, although it is wise to check this with the particular machine. In addition, there should be no uprights that might interfere with discharge from the feeding wagon.

Keeping cows out of the feed

There is no easy answer to this problem as it depends on the feeding system. Some ideas include:

- To reduce bullying, cows can be separated into smaller groups, ensuring there is sufficient room at the feeding face.
- The trough length should be no more than 70 cm/cow if they are all feeding at once or 30 cm/cow if they are fed *ad lib*.
- Ensure the inside floor of the trough is 10–15 cm higher than the feedpad surface to provide easier access to the feed.
- Install hot wires down the centre of the trough.
- Install pipe head bales. Although expensive, they prevent cows from climbing into or being pushed into troughs and also prevent them from lifting their heads out of the troughs hence throwing feed around.

Other aspects of feedbunk management

The term 'feedbunk' is traditionally used in the US for feeding strips, but the following practices also apply to trough feeding. To maximise feed intakes and milk production in lofted stock, feedbunks should not be left empty for more than 2 or 3 hr/d. When feeding time is limited to less than 8 hr/d, milk production can be reduced by 5–7% in mid lactation cows, and to an even greater extent in high producing cows, which are at or near peak lactation; push up the feed regularly to encourage feeding and minimise sorting of ingredients; the feed barrier should not restrict cows' reach; it is important to provide sufficient feedbunk space so competition does not adversely affect feed intakes but not too much space so stock can throw it around as they eat. As first calf heifers tend to eat smaller meals than older cows, separating them out in large herds might reduce competition and improve their performance. The feed surface is ideally 100 mm above feet level. The feed area should be under cover to prevent rain decreasing feed intakes. Remove residue fed and use for other groups, such as lower yielding or dry cows. Cattle consume most of their feed during the comfortable period of the day; that is, in late evening in hot weather and during the middle of the day in cold weather. That is the time they should be offered the most.

Cows are animals of habit that like routine, so once a schedule has been developed, stick to it and if changes are necessary, allow time for the cows to adapt and monitor feed intakes to decide if it was worthwhile. Ideally, feed residues should be restricted to 3% or less without impacting on milk production or cow health.

US researchers have developed a scoring system to quantify feed residues from complete or partial mixed rations (Bolsen and Pollard 2004). This provides a detailed description of the feed remaining (see Table 8.1) which can be assessed over a four day period to help monitor cattle responses to changes in feeding procedures.

Table 8.1 Suggested scoring system to quantify feed residues.

Score	Description
0	No feed remaining
0.5	Only scattered feed remaining from previous feeding. Most of feed trough/strip is exposed
1	Thin, uniform layer of feed remaining across bottom, typically less than 1 cm thick
2	25–50% feed remaining from previous feeding
3	The crown of previous feed is thoroughly disturbed but more than 50% of feed remains
4	The crown from previous feed is still noticeable and feed is virtually untouched

(Source: Bolsen and Pollard 2004)

Integrating feedpads with grazed pastures

Feeding management of feedpads should be aimed to fill in the feed gap resulting from shortfalls in pasture. Homegrown pastures or grazed crops are likely to be of higher quality and lower costs than purchased forages so should be utilised first. High fibre by-products can also be purchased to extend farm fodder reserves, but these are unlikely to produce as much milk as cereal grain and other high energy concentrates.

Drying off earlier and culling cows can extend pasture reserves but will have a future impact on farm finances and long-term farm business financial position. Non-lactating stock still have high nutrient demands; from Table 6.1, a dry cow still requires 125 MJ of ME/d in her ninth month of pregnancy. Agisting dry or young stock can increase pasture supplies for milking cows.

Following rain, bare paddocks should be sown down to opportunity crops, even just to provide short-term feed.

Principles of effective grazing management

Detailed discussions on the principles of grazing management are beyond the scope of this book, but the key principles should be understood to maximise the benefits of a feedpad in maintaining the most profitable grazing practices on pasture based dairy farms. These are as follows:

- Pastures should be grazed at their optimal quality, or stage of pasture growth. This is largely determined by the rotation length between grazings which varies with the rate of plant growth, largely a function of season.
- Pasture quality can be monitored by counting the number of new leaves on each plant and grazing when they reach say, 2.5–3 leaves in perennial rye grass pastures.
- Forage conservation of selected paddocks is an integral part of grazing management during the spring flush of pasture growth, to maintain the most effective grazing rotation.
- Sufficient time must be given between fertilising and grazing pastures for pastures to respond to the improved soil-nutrient status. In spring for example, four to six weeks is ideal whereas beyond eight weeks, pasture quality will rapidly decline.
- Post-grazed pasture profiles should be carefully monitored to assess whether the pasture has been undergrazed, grazed just right or overgrazed. For example, what % of the pasture is made up of clumps of ungrazed pasture? Is the pasture contaminated by dung and urine or just too mature to be selected by grazing stock. Have the cows eaten into the sides of these clumps? Have they eaten the tops out of these clumps?

- What is the post grazing height in between these clumps? It should be 4–6 cm for temperate ryegrass pastures.
- Post-grazed pasture characteristics can indicate whether there has been too much substitution of pasture for feedpad supplements.
- Concentrate residues – leftover concentrates in the milking shed or the feed trough can indicate palatability problems, too much is being fed during the milking period or the feed system is not supplying the correct amount to every bail.
- Feedpads will reduce soil and pasture damage, thus leading to faster winter and spring regrowth. This extra pasture should be consumed or it will deteriorate thus reducing its palatability.

There are also many simple observations farmers can use to highlight potential feeding problems in grazing cows when formulating and feeding out rations on the feedpad. These have been discussed in the section 'Troubleshooting potential feeding problems' in the previous chapter.

Maintaining pasture quality can be achieved by increasing stocking rate; calving earlier; identifying pasture surpluses quickly and making more silage of higher quality earlier; reducing supplementary feeding earlier in the season; and observing pasture residuals post-grazing carefully and acting accordingly.

Using computers to aid feeding decisions

To quantify the nutrient intakes of grazing stock supplemented on a feedpad, we firstly need to know the amount of pasture consumed and its energy, protein and fibre contents. As there are many factors influencing how cows select grazed pasture, this is not easy. At least for Victoria, a computer program has been developed to assist with these predictions. This program called DIETCHECK, (Heard *et al.* 2004, Jenkin and Wales 2007) predicts the amount of pasture and nutrients consumed by milking cows that strip graze pasture, up to 50% of their total daily energy intakes, with or without supplements.

The operator selects the region in Victoria, terrain (flat, undulating or steep) and month of year; dominant pasture species (ryegrass, clover, other grasses) and pasture height; herd size and pasture area; average level of production (milk yield and composition, live weight, stage of pregnancy, body condition and change in condition); type and amount of supplement; and milk price and cost of each supplement.

The program then predicts the energy, protein and fibre contents of the selected pasture; calculates the rate of substitution of pasture for supplement, which can be accepted or altered; predicts grazed pasture intake, total intakes of DM, energy, protein and fibre and the balance of each nutrient (that is deficiency or excess); marginal milk production (per kg supplement DM fed); daily milk return and cost of supplements (cents/d); and financial benefits (expressed as cents/kg supplement DM fed).

To manipulate the predicted marginal milk responses and financial benefits, the user simply manipulates the management factors over which he has most control on the farm, namely the pasture height, area of grazed pasture, supplement feeding regime and supplement cost.

The program utilises data on selected pasture from Victorian temperate pastures and does not consider environment stress and other constraints to pasture intake and milk responses. Its major limitation for use in pasture-based systems incorporating feedpad technology is that at least 50% of the energy intake must be derived from grazed pastures. It works well when supplement levels are less than 8 kg DM/cow/d.

Other computer programs are available to aid feeding decisions, but DIETCHECK is one of the very few developed specifically to predict nutrient intakes from grazed pastures, a prerequisite for accurately formulating rations for stock supplemented from feedpads.

Dairy Australia (2009d) recently compared the predictive ability of 11 different dairy nutrition computer models, currently in use in Australia under a variety of feeding scenarios. With regards predicted marginal responses to increasing levels of supplement, all but two assumed linearity, which from the section 'Deceasing marginal responses' in Chapter 6, is highly unlikely over the range of feeding levels encountered in Australia. Of even greater concern was the inability of any of the models to accurately predict the likely milk responses to grain feeding. With one exception, predicted milk responses in all 11 models were 2.2 to 2.8 kg milk/kg triticale, which from the section 'Immediate and delayed milk responses' in Chapter 6, would over-predict them by 200 to 400% in grazing cows. Clearly, computers may be an aid in feeding management, but cannot replace practical experience when planning PMR feeding programs with grazing dairy cows.

Checklists for feeding management

To optimise cow performance and minimise potential problems with cow health, the following checklists should be followed:

Planning the diet

- Insufficient energy and protein in ration will reduce milk yields.
- Insufficient effective fibre in the diet can lead to acidosis. Diets with less than 30% NDF are high-risk.
- 75% of the fibre sources should be greater than 1.5 cm in length.
- Feed forages before feeding grain or energy-rich concentrates in the dairy shed.

- If you cannot feed forages first, make sure cows get access to them quickly after they leave the milking shed.
- Aim to feed the best available quality forages to the milking cows, particularly those in early lactation with the highest milk yields.

Sourcing feed ingredients

- Spoilage or contamination of feed ingredients can reduce cow performance.
- Mycotoxins in hay, silage and other wet feeds can cause abortions.
- Excess nitrates in some forages can cause nitrate poisoning.
- Dead rodents in feed ingredients can cause botulism.
- Ensure silage bunkers have clean sheared face with no residues or moulds and have smallest area possible unsheeted at any time.
- Plan feed purchases carefully to reduce costs and storage requirements.
- Feed ingredients should be tested to ensure a better balanced diet.

Measuring and mixing feed

- Incorrect measuring of ration ingredients can increase risks of acidosis and reduce milk yields.
- Feeds with lighter and larger particles (forages) should be loaded before those with heavier and smaller particles (concentrates).
- The ration density should be light and fluffy, not wet and lumpy.
- Poor mixing can lead to feed sorting so some cows consume too little fibre and increase acidosis risks.
- Poor mixing can also lead to some cows eating lumps of ingredients, such as urea, which can have disastrous effects on cow health.
- In herds calving year-round, all cows can be safely fed the same ration on the feedpad with additional grain fed to high-yielding cows in the milking shed.
- Rations containing less than 45–50% DM may depress intakes. There will be some seasonal variability in ration DM% as ingredients alter from one month to the next.
- Ideally, feed 5–10% more than cow's daily requirements.
- Uniformity of mixing is essential to ensure consistent particle sizes in the mixed ration.
- Particle size can vary with condition of mixer wagon, quantity of feed mixed each load, loading order, type of feed and particle size prior to mixing.
- Clean out wagon every day.
- All ration changes should be made at the afternoon feeding to eliminate the possibility of feeding hungry cattle a new, higher energy diet in the morning, which could increase digestive upsets such as rumen acidosis.

Delivering the feed

- Dirty or overcrowded feed areas will reduce feed intakes
- Cows spending more than 4–5 hours in mud and slurry can increase foot rot.
- Teat end contact with manure within 30 min of milking can increase mastitis.
- Crowded stock push each other thereby increasing risk of lameness.
- Spread of diseases is greater when cows spend time in small areas.

Grazing management

- Ensure pastures are grazed at their optimum quality, such as 2.5–3 leaf stage in perennial rye grass.
- Post-grazed pasture characteristics can indicate whether there has been too much substitution of pasture for feedpad supplements.
- Sufficient time must be given between fertilising and grazing pastures for pastures to respond to the improved soil nutrient status.
- Conserve excess spring pasture growth to maintain rotation lengths to optimise pasture quality.

Feeding management during periods of heat stress

- Cows normally drink 120–150 L/day but this can increase to 200–250 L/day during hot periods. A big water trough should be located on the exit side of the dairy.
- Water pipes should be 75 mm in diameter with sufficient pressure to provide 20 L/cow/hr so troughs can cope with peak demands.
- Feed a nutritionally balanced, high-energy ration during hot periods, because high-fibre rations can increase internal body heat production.
- Consider increasing levels of sodium, chlorine and potassium in the diet as these are lost in sweat during hot periods.
- Feed earlier in the day, even before 0900 hr on days of high heat stress.
- If planning to graze the cows, provide the pasture during the evenings when cows can cool off and their appetites will be higher.
- Professional advice should be sought to formulate the best rations for high-yielding cows during hot periods.
- See Chapter 9 for additional advice on heat stress management.

9

Cow management

This chapter discusses the key issues of ensuring animal well-being while on feedpads such as cow welfare, animal health and environmental management.

The main points in this chapter:

- As animal welfare issues are attracting increasing scrutiny, stock on feedpads require closer attention than while at pasture.
- The key animal health issues for cows on feedpads are mastitis, lameness and acidosis. Lameness can be monitored easily in stock, using a locomotion score.
- The physical state of a cow is a good guide to its health status. Farmers can do a lot to assist veterinarians in diagnosing and treating animal health problems.
- Heat stress can become a major problem throughout Australia's dairy regions. The best single descriptor is the Temperature Humidity Index as this combines temperature and relative humidity into a single comfort index.
- The thermoneutral zone of a milking cow is between 6–18°C. Outside this range, cows must modify their metabolism to remain comfortable. Above 27°C, appetite is depressed and both biological and economic efficiencies decline.
- Monitoring respiration rates provides a good indicator of heat stress. As well as feedpad design, manipulating feeding and herd management strategies can reduce heat loads.
- The effects of low air temperatures, wind, wet hair coat and mud are additive when assessing cold stress. The lower critical temperature can vary from -30 to +13°C in dairy stock depending on their physiological state.

This chapter discusses the key issues of ensuring animal well-being while on feedpads such as cow welfare, animal health and environmental management (see Figure 9.1).

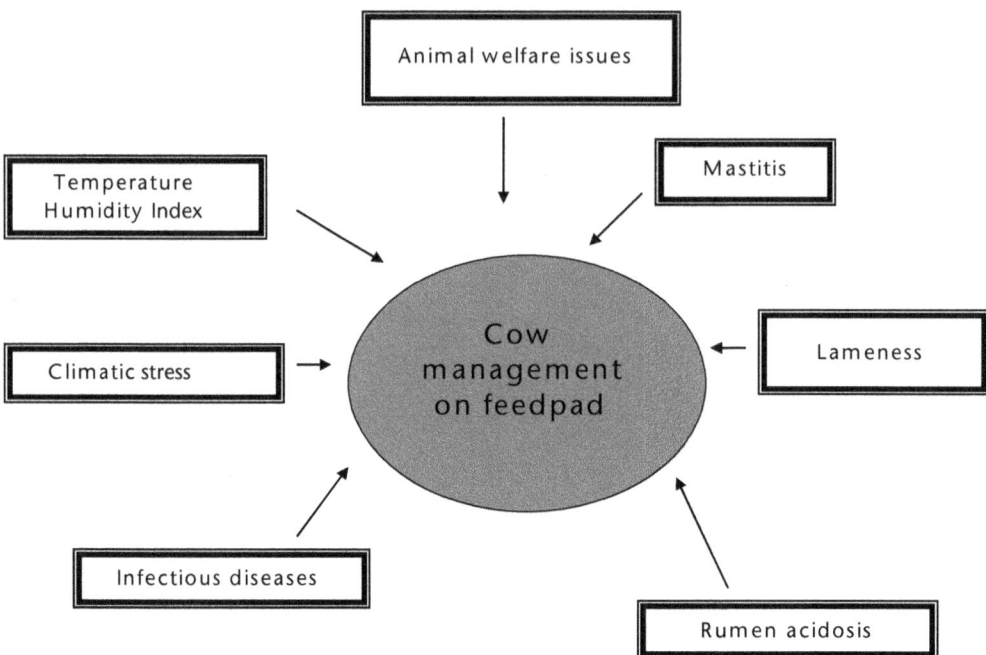

Figure 9.1 Some of the key considerations for good cow management.

Animal welfare issues

Under the animal welfare legislations of all states, those who have the care or charge of stock have a 'legal duty of care' and a moral responsibility for their welfare. As well as the owner, this includes farm employees, as their designated representatives. Such animal welfare standards are being used by the courts as yardsticks to assess husbandry and management practices in animal cruelty cases.

The basic requirements for welfare of cattle are:

- water, food and air to sustain health and vitality;
- social contact with other animals;
- sufficient space to stand, lie down, stretch and groom;
- protection from predators;
- protection from disease;
- protection from adverse effects of extreme of climate or unseasonal changes in weather conditions;
- provision of reasonable precautions against the effects of natural disaster, such as providing fire breaks; and
- protection from unnecessary, unreasonable and unjustifiable pain, injury and suffering.

The importance of good stockmanship in animal welfare cannot be overemphasised. Good stockmen have flexibility in their approach to stock handling and adapt to the needs of differing stock circumstances.

Animal welfare issues are attracting increasing scrutiny, particularly when intensifying existing livestock farming operations (Chamberlain 2002). In this context, managers of dairy feedpads need to provide well-ventilated air; adequate sources of clean water; adequate space for offering high quality feed; non-slippery and even walking surfaces; clean and dry resting surfaces; adequate space for stock to ruminate and rest; calm, quiet and non-stressful stock handling; and shade and shelter for stock throughout the year, so as not to magnify heat or cold stresses. They should minimise the opportunities for aggressive stock behaviour, such as between dominant and submissive cows, that can affect overall cow performance adversely. They should ensure feedpad cleanliness, as stock are restricted in their ability to find a clean and dry place on which to lie. They should be able to quickly address any animal health issues because of the increased exposure to disease agents, and they should also provide any additional stockman skills required to manage confined stock for lengthy periods.

The welfare of stock can be determined by their capacity to avoid suffering and sustain fitness. Suffering occurs when animals fail to cope with stresses because they are too severe, too complex or too prolonged. Some key indicators of poor welfare can include abnormal behaviour such as lack of rumination, cows hanging their head, excessive stiffness and lameness; unusual resting behaviour such as cows at pasture choosing to lie down rather than graze, or cows choosing not to lie down in the feedpad after a period of grazing and cows lying in alleys in free stall sheds; lowered disease resistance as evident from increases in the incidence and severity of any disease; traumatic injuries and body malfunction; reduced performance (growth rate, milk production, fertility), lowered reproductive ability, and shorter life expectancy. Look also for increases in the frequency that stock must deal with situations which they normally strive to avoid; increasing incidences when stock are deprived of their preferred food, water, sunlight and shelter which can affecting their ability to carry out normal behaviour adversely; obvious reductions in their degree of contentment as indicated by any abnormal behaviour; deleterious changes in appearance and condition, and changes in the stock's ability to cope with the more intensive management encountered on feedpads.

Stockmen working with confined stock need to be aware of the skills for managing and handling such cattle, by minimising any stress; using the natural behaviour of cattle; recognising the early signs of distress or disease and initiating prompt and appropriate preventative and remedial action. They should also be aware of the 'fight or flight' response of stock and the 'threat behaviour' in stock (see below).

Fight or flight behaviour. Agitated or alarmed animals show characteristic behaviour which can include ears up, eyes bulging and tail tucked under. Such an animal will dislike being close to stockmen and will probably be looking for an escape route. It could either try to climb through or over apparently solid structures or may even turn on the handler. This is often more apparent in bulls with a group of cows or cows still with their calves.

Threat behaviour. Even though dairy cows are domesticated, they can still show behaviours that evolved from their primitive ancestry. For example, an animal facing the threat (such as a stockman) with head lowered, tail swishing and maybe even bellowing, provides a warning that can quickly be followed by a full attack. The person should start to back away slowly but not turn their back to such an animal. If it starts to charge, then the person should outstretch his arms to make himself look bigger while yelling loudly and menacingly. By taking such advantage of the animal reassessing the situation, they should quickly move to a safer location.

Most dairy farmers are innately aware of many of the signs exhibited by stock when they become stressed. Furthermore, farmers generally have close working relationships with their professional dairy service providers. It is important to discuss stock welfare and health management quite specifically with their advisers since the lengthy and close confinement of cows on a feedpad can quickly magnify potential disease problems.

Animal health issues

The greatest concern about feedpads is the perception that stock are concentrated in an environment considered to be uncomfortable and unhygienic. This could well be the case on poorly designed and managed pads where manure builds up the level of pathogens. On well-constructed and maintained feedpads, exposure to such disease-causing organisms is likely to be lower than on muddy paddocks.

A professionally developed animal health plan is essential in any large-scale dairy operation. Such a document should be readily available and easily accessible to all staff. It should include acceptable (and unacceptable) animal husbandry and management practices; relevant Codes of Practice of animal husbandry; priority cases for veterinary attention, such as cows that cannot stand or walk unaided following calving, those with hip and leg injuries, and those that do not respond to routine treatments; vaccination and parasite control procedures and records; biosecurity procedures and contingency plans for disease outbreaks; specific sections on managing lameness, mastitis, acidosis, trauma injuries and other conditions associated with intensive dairy production systems; euthanasia procedures and appropriate carcass disposal; and the necessary staff resources and training procedures.

Theoretically, infectious diseases could increase due to cow concentration on a feedpad and a potential build-up of infectious organisms. As cows are likely to be better fed and with less climatic stresses than at pasture, greater cow comfort and reduced nutritional stress should improve their disease resistance.

Good management of feedpad effluent is essential as raw manure and recycled effluent contain organisms responsible for conditions such as Johne's Disease, salmonellosis, leptospirosis, mastitis, coccidiosis, clostridia, Enzootic Bovine Leucosis and internal parasites. Practices to reduce such risks include not grazing young stock on pasture that has received raw manure or recycled effluent, not grazing any stock for three weeks following manure or effluent applications, or until the pastures have received at least one week's strong sunlight; consider botulism vaccination if fertilising pastures with chicken manure; consider a north–south orientation for covered feedpads, to allow morning and afternoon sunlight to dry cow standing areas; compost manure if reusing it for bedding; undertake effective fly control around discarded or spoilt feed; undertake effective control of vermin and wildlife, particularly around feed storage and preparation areas; and routinely dehorn all replacement heifer calves.

Lameness

When feedpads are used to keep cows off pastures for the entire grazing period, they will reduce their walking distances for the day. Problems with foot bruising, worn soles and foot abscesses, often caused by walking on tracks, can also decrease. Lameness problems in Europe and the US caused by laminitis are often due to lack of exercise. This is unlikely on the vast majority of farms in Australia and any country where a feedpad is used in combination with grazed pastures.

Potential lameness problems will depend on the surface of and stocking density in the feedpad. Hard surfaces such as concrete are not recommended for cows standing over lengthy periods as this can increase lameness, stiffness and agitated behaviour, all of which affect cow performance adversely. Lounging areas are required if stock stand off to protect pasture and soils and these should provide a soft, clean, non-slippery surface for cows to lie down.

If cows are destined to spend long periods on concrete, they should be selected for superior leg and feet conformation. Selection should be based on hooves that are short, angled steeply, with high heels and even claws. The sole should be concave with the majority of the weight taken by the hoof wall. Poor conformation should be avoided where possible, including overly straight hocks, weak pasterns, sickle hocks, splay or overlapping toes, as these may increase the rate of lameness. Some hoof conformation may be restored by trimming.

The trigger point to look more closely at potential lameness problems is if the herd contains more than five lame cows per 100 cows at any one time. With higher incidences, cow crowding at the feed face should be reduced by splitting the herd into small groups for feeding. If cows have to stand for more than four or five hours in mud and slurry, footrot and other feet infections are likely to increase. Lameness is discussed in more detail later in this chapter.

Mastitis

The concentration of mastitis-causing organisms will be minimal if the feedpad is regularly cleaned properly. The amount of mud must be reduced and the pad should have adequate bedding and drainage.

Any feeding area used for more than 2 hr per day will increase the chances of teat-end faeces contact, leading to more environmental bacterial contributing to increasing cases of clinical mastitis. The degree of teat end contamination can be reduced firstly, by not allowing cows to loaf on the feeding area and secondly, by having a regular program of scraping and cleaning the feedpad surface. The threshold for more active prevention of new mastitis infections is:

- five clinical cases for every 100 cows during the first month after calving, and
- three clinical cases for every 100 cows per month for the remainder of lactation.

Cows should be selected for good udder and teat conformation. They should have compact and non pendulous udders, short teats with healthy skin and good teat sphincters that allow free milking but close quickly following milking. Replacement heifers should be selected from cows with a low history of mastitis while bulls chosen for AI should also have a history of offspring with minimal clinical mastitis.

If feedpad hygiene is sub-optimum, cows should be discouraged to lie down until at least 30 min after milking, to provide sufficient time for the teat canal to close post-milking. Feeding cows immediately on their return from the milking shed should encourage them to stand long enough.

Teat condition in cows will always deteriorate during wet weather due to increase in soiling of teats and the teats being wet for lengthy periods. Windy conditions can also cause teat chaffing. For this reason, teats may be seen to deteriorate when cows are on a feedpad during wet periods, but this is likely to be due to the weather and not the feedpad itself. As the feedpad will keep cows away from muddy pastures, if kept clean, teat condition should then improve.

Acidosis

Since feedpads provide for a better forage feeding system than paddock feeding, balancing the diet with adequate fibre for normal rumen function and good milk composition, should reduce rather than increase potential problems of acidosis and other metabolic diseases. As a rule of thumb, if pasture DM intake is less than concentrate DM intake, dietary fibre may be limiting, thus requiring supplementation with conserved forages. Excessive rapidly fermentable carbohydrates in the diet can lead to clinical and sub-clinical laminitis with symptoms similar to lameness.

It is important that all stock have access to and actually eat the available fibre in the ration. Having adequate trough space per cow should ensure that the dominant feeders do not eat more than their fair share. Ensuring the ration is mixed properly will result in the concentrate not settling out prior to feeding. Rumen buffers and other feed additives play a role and are discussed in Chapter 7.

Fertility

The flexibility of feedpads to combat winter feed deficits can ensure sufficient energy intakes and better overall body condition. Improved energy intakes during late pregnancy and early lactation will improve body condition hence reproductive performance. Cows should cycle sooner after calving thus get back into calf more quickly.

Detecting heats should also be easier since herd behaviour can be more closely monitored while on the feedpad, saving a trip down to the paddock for heat detection. Slippery pad surfaces, however, may reduce the incidence of cows mounting each other and a lot of bulling activity on slippery surfaces can cause hip injury. More sensitive indicators such as tail paint or heat mount detectors should be considered when cows are jumping less while on the feedpad. In addition, attention could be given to ensuring feedpad surfaces do not restrict such natural mating activities.

The feedpad complex should have appropriate facilities and procedures for mating and calving. Accurate records of oestrus and mating are essential to predict calving dates hence observe cows close to parturition.

Physical attributes of healthy and sick cows

It is not easy to say what 'disease' means. In a general sense it means anything which is not 'normal'. Disease is then a condition which is detrimental to the health and well-being of that animal. It can include injuries, infections by micro-organisms, infestations of parasites, nutritional deficiencies, poisoning and hereditary abnormalities. The presence of disease, acute or chronic, reduces cow performance which invariably reduces farm production and profitability.

The physical state of a cow is a good guide to its health status. Healthy animals are alert, active, have bright eyes, with no discharge, smooth and shiny skin. They breathe and urinate regularly and their tail moves to drive flies away. Signs of stress include loss of appetite, reduced daily milk yield, increased temperature, high respiratory rate, tongue protruding, open-mouth breathing, inability to lie down and a preference for remaining in cool waters; for example, farm dams and watercourses.

Symptoms of health problems

These can vary from one extreme, almost unnoticeable changes in behaviour, to the other extreme, death. Symptoms of health problems are many and varied. They include:

- *Nutritional status*: Cows can be fat, normal or thin. Thin cows are not necessarily sick, as they could have recently calved or simply be high producing animals. Sick cows tend to lose weight due to depressed appetite, poor feed digestion or loss of body reserves. Cow condition should then be judged in relation to all circumstances.
- *Walking and standing*: The way an animal moves can indicate pain in the body, the result of a traumatic injury or an infected hoof. A method of scoring the cow's ability to walk (locomotion score) which provides a good guide to lameness, is described later in this chapter.
- *Eyes and ears*: Eyes have a bright and lively expression with no discharge; sunken eyes indicate dehydration. Ears should be able to move around freely.
- *Skin, coat or mucous membranes*: The skin of healthy cows is flexible and when pinched, should quickly return to normal; a lengthy delay will indicate dehydration, as will a dry nose. The coat should be smooth and shiny. The mucous membranes around the eye, nose and vagina should be pink to reddish in colour and be moist. In sick cows, these membranes can become either too red or too pale, the latter indicating anaemia.
- *Digestion:* Healthy cows have a good appetite and eat with eagerness. Faeces and urine are discharged regularly with the faeces having a normal consistency. When digestion is disturbed, the cow's appetite decreases and the faeces is discharged too fast (scours) or too slow (constipation). Cows ruminate frequently when healthy (at least 6–8 hr each day), and if they do not ruminate when resting, their digestion is disturbed. Trouble-shooting problems with feeding management has been discussed in Chapter 7.
- *Urine*: It should be thin, yellow and clear; thick, mucous or red urine is an indication of ill health.
- *Vagina*: It should be closed, with no swelling, no discharge and be slight whitish-red in colour; a swollen vagina with whitish discharge or decomposing membranes is indicative of reproductive problems.
- *Respiration*: In healthy cows, respiration is quiet and regular, whereas in cases of unrest, fever, fatigue or heat stress, respiration rates increase. See a following section in this chapter for a guide to heat stress based on respiration rates.

Coughing, nasal discharge, rapid or slow breathing can all be symptoms of ill health.

- *Blood circulation*: Pulse rate is a good guide to cow health, with 60–70/min being normal for dairy cows.
- *Body temperature*: Normal body temperature is 38.5–39.5°C. Higher temperatures are recorded in sick animals (with a fever) or in heat-stressed animals.
- *Milk production*: When a cow is sick, milk production drops, due primarily to decreased appetite.
- *Specific signs to look for in calves*: Health problems are likely in calves with droopy ears, head down, not drinking, lying in a corner, dribbling, limping, swollen joints, swollen navel, scour or blood in their faeces. The nose and eyes should be clear and damp with no discharge. Calves should stretch when they stand up following a rest period. An odour of ammonia can indicate poor ventilation and potential pneumonia problems.

Other tools to diagnose health problems

There is much that farmers can do to assist veterinarians or other animal health professionals in diagnosing and treating such problems. These include:

- Examine the history. For how long has the animal been sick, how rapidly did the sickness develop, what are the breeding and calving records of the animal, have new stock recently been introduced to the farm, have there been any recent environmental changes (such as hot weather)?
- Examine the animal. Use the checklist above.
- Examine the environment. Are other animals sick, could there be poisons or mineral deficiencies involved, what is the water quantity and quality, what are the physical conditions, such as lanes or yards, that cause traumatic problems?
- Prepare for the veterinary visit. Record the important major symptoms using the above checklists, have the sick animal easily accessible with suitable restraining equipment available, listen carefully and take notes for follow-up treatment.
- Once the disease has been diagnosed and its cause, symptoms and treatment identified, it is important to develop a control program to reduce its incidence in the future. Veterinarians can assist with such a control program. For the farmer, prevention is just as and often more important than cure.

Monitoring stock in free stall sheds

All staff members should be well trained in observing normal behaviour and recognising and reporting abnormal or unusual behaviour and other signs of emerging health and welfare issues. The following observations can assist with monitoring stock in free stall sheds:

- Do cows appear comfortable when standing or lying?
- Do cows lie backward in stalls or alleys?
- Do cows stand half in or half out of stalls?
- Do cows stand in stalls in an angular fashion?
- Are all stalls used equally?
- When cows normally rest (say between 1000 hr and 1600 hr), are more than 20–30% of the herd standing in the stalls?
- Are cows udders dirty? Also look for dirty tails and hindquarters.
- Check for patches of rubbed-off hair and injuries to the hocks and knees, as this may indicate cows rub excessively on stall partitions or neck rails when standing or lying down.
- Are cows walking very slowly, or timidly, with rear feet spread apart, indicating poor traction or laminitis?
- Cows bellowing excessively or abnormally can be an indicator of inappropriate handling or facilities.
- Are there any visible injuries or other lesions?
- If more than 20% of the cows defecate in the milking yards and shed, this could be a sign of discomfort or uneasiness.
- Use the following tests (O'Keefe *et al.* 2010) to indicate bedding comfort:
  - Wet knees test – kneel in the stall for 10 seconds. If the knee is wet, then the stall bedding is not dry enough.
  - Drop knee test – crouch and then drop to your knees in the stall. This will quickly tell how truly comfortable the stalls are for the cows.
  - An impact soil tester can be used to compare the compressibility of different surfaces and any changes over time, which correlates well with cow preferences.

Identifying, treating and preventing lameness

Lameness is very often due to bruised soles causing pain within the claw. Bruised soles result from excessive pressure on uneven ground, especially on stones. Inside the hard, outer layer of the hoof wall and sole, there is a sensitive layer rich in blood vessels and nerves. If a cow stands on a stone, or some other small hard object, its sole bends upwards over the stone, severely squeezing the sensitive layer. This can cause bleeding within the claw, and subsequently pressure, pain and lameness.

Bruising is identified in a well-cleaned sole as pink or dark red flecks. Very soft feet, due to moist or wet conditions are more prone to bruising.

Small sharp stones lodged between the claws can also cause discomfort and lameness. They can also perforate the skin and act as a point of entry for bacteria causing footrot. On hard ground, they can puncture the sole leading to sole abscesses. Sharp edges of broken concrete can also cut or bruise soles.

Multiple causes of bruised soles

A number of factors contribute to increased risk of bruising. The more stones on tracks or in yards, the greater the risk of bruising. Damages are more likely if the stones are free on the surface rather than buried, if they are sharp rather than rounded and if they are on a hard surface, such as concrete. If a track breaks up, and stones from the track base are brought to the surface, there is increased risk of bruised soles. If the hoof is soft due to very moist or wet conditions, it distorts more easily over a stone and offers less protection to the sensitive layer, hence is more prone to bruising. If the sole is thin, due to excessive wear, it will offer less protection. Such wear can result from standing on very rough concrete floors or animals being bullied by dominant animals or even excessive turning on floors when on heat. It can also result in heifers from bullying by older and more dominant cows. Driving or pushing cows along a track or into an already crowded yard can result in a cow lifting her head and placing her feet in an unplanned manner. She is then more likely to step on a stone and cause bruising. Poor hoof shape or conformation can use excess hoof under the sole. Then the normal pressure of walking may be sufficient to cause bruising and lameness. Multiple causes can be present at the one time. For example, winter is a common time for excessive moisture, and for tracks to break up, which in turn may result in a herd being driven too fast. Thin soles may be also be present in young stock.

Assessing cow lameness

Lameness is an increasing problem in both grazing and housed cows, with economic implications. Locomotion scoring from 1 to 5 (for increasing lameness) is a new tool (Sprecher *et al.* 1997), which provides a quick measure of the cow's ability to walk normally. These are presented in Table 9.1. Observations should be made of cows' standing and walking (gait), with emphasis on their back posture. They should be made on a flat surface that provides good footing for cows.

Locomotion scores of individual cows can be used to select cows for hoof examination before they become clinically lame. Those with scores of 2 and 3 are considered sub-clinically lame and their hoofs should be examined and trimmed to prevent more serious problems. Scores of 4 and 5 represent those cows clinically lame. The higher the lameness score, the greater the reduction in feed intake and milk yield and the poorer the body condition. For example, a score of 4 can reduce DM intakes by 7% and milk yields by 17%, while a score of 5 can reduce DM intakes by 16% and milk yields by 36%. Advice should be sought if more than 3% of first calving cows, or more than 2% of older cows, show signs of lameness.

When examining individual cows closely, look at the way she stands and examine how she walks (see Table 9.1). Look for obvious signs of swelling or muscle

Table 9.1 Locomotion score guide based on observations of back posture and behaviour when walking.

Score	Clinical description	Back posture	Assessment
1	Normal	Flat	Cow stands and walks with a level back. Gait is normal.
2	Mildly lame	Flat or arch	Cow stands with level back, but arches when walks. Gait is slightly abnormal.
3	Moderately lame	Arch	Stands and walks with arched back. Short strides with one or more legs.
4	Lame	Arch	Arched back is always evident and gait is one deliberate step at a time. Cow favours one or more legs/feet but can still bear some weight on them.
5	Severely lame	3-legged	Cow demonstrates an inability or extreme reluctance to bear weight on one or more limbs/feet.

(Source: Sprecher *et al.* 1997)

wastage and check for wounds in the skin and discharge. If the cause of lameness is not obvious, lift the lame leg and clean the foot thoroughly ensuring the two claws and the space between them are clean. Check this space and explore the tight area between the heels with your finger for foreign objects. Also check for unusual growths, blisters or vesicles. Scrape or sand the entire surface of both heels to check for cracks, bruises and ulcers, using a hoof knife or angle grinder with course paper. If no ambiguity is found, more closely check the sole of both claws very carefully. If no abnormality is still apparent, examine the rest of the foot for evidence of abscesses. If no abnormality is still apparent, examine the rest of the leg for wounds or swellings. If no obvious problems are apparent, seek professional help.

Treating cow lameness

Draft cows with bruised soles into a paddock close to the shed at milking time. Bruising will repair with time, but rest is important. Walking long distances can lead to additional bruising. Make sure cows don't have to walk long distances for food or milking.

Particular severe bruising may need some form of relief from the pressure of body weight and walking. If bruising is largely confined to one claw, glue-on plastic or leather lace-up shoes can be fitted. Preventative measures include ensuring floors are not too abrasive, with all stones and broken pieces of concrete removed. Hoof trimming will also assist. Foot-baths containing formalin (for hardening hooves) or sprays of zinc sulphate solution (for treating sore feet) are also useful.

Preventing bruising

Several approaches should be used. Reduce stones with appropriate track maintenance. Repair unstable areas and properly construct and maintain tracks.

Reduce moisture by draining muddy areas and ensure good drainage on feedpads. Ensure alleyways in free stall sheds are regularly cleaned so cows are not walking in faeces and mud continuously. Rubber mats may prove useful in reducing hoof wear in high-pressure yard areas. Avoid conditions causing excessive sole wear, such as removing abrasive sand from yards and pad surfaces. Ensure good stockmanship when moving cows. Keep hooves in good shape with corrective trimming if necessary. Hooves can be trimmed at drying off when cows have a rest from walking long distances. Seek professional assistance to develop a hoof health plan which includes training, monitoring, recording, prevention, diagnosis and treatment.

Preventing other physical trauma injuries

Conditions of the upper limb, such as hock lesions, knee lesions and adventitious bursa may also become more pronounced in cattle restricted to feedpads. These trauma injuries are often due to inappropriate bedding type or lack of thickness. Hip and shoulder injuries are indicative of poor handling and poor post and rail placement. Gateways, fencing and free stalls should be maintained to minimise projections, such as broken boards or rails, or protruding nails.

Minimising heat stress in dairy herds

The comfort zone for milking Friesian cows is between 6°C and 18°C. Within this range, there are no measurable fluctuations in their physiological processes while the energy input to output shows good biological efficiency, in that all body processes will be functioning in their expected ranges. Between –5 and +5°C, appetite will be stimulated while at the upper level, above 27°C, appetite is depressed and both biological and economic efficiencies decline. Above 24°C, feed intakes decreases by about 3% for every rise of 1.2°C.

The impact of high temperatures on cow appetite depends on the type and quality of forage; intakes of high fibre forages are more depressed at high temperatures; type and quality of concentrates; cows will choose to eat more concentrates and less forages when heat stressed; humidity – high humidity exaggerates the effect of high temperature; stage of lactation – cows in early lactation are more susceptible to heat stress; milk yield – high yielding cows are more susceptible to heat stress; breed – Friesians are more susceptible than Jerseys; actual appetite – cows with large appetites are the more susceptible to heat stress.

When planning feeding programs, consideration should be given to the number of hours each day when temperatures exceed 27°C and relative humidity exceeds 80%; feed intakes will decline once temperatures exceed 27°C or higher for six hours. Furthermore, high body temperatures reduce the efficiency of rumen digestion and increase body maintenance requirements, further increasing the energy deficit. The net effect on feed intake also depends on the number of hours each day below 20°C, which allows cows to cool hence restore their heat balance.

High temperatures can increase growth rates of pastures, but associated with this is often an increase in fibre and decrease in energy levels in the forages. Reduced forage quality and imbalances in feed nutrients invariably reduce milk yields and increase dry periods, due to delayed oestrus. Lengthy dry periods, exceeding 120 days, can produce over-fat cows, increasing the likelihood of metabolic diseases following calving. In warm climates, young stock will use 15% more energy to maintain heat balance than will their peers in cooler climates. High humidity levels can also lead to higher incidence of respiratory problems, with long-term effects on cow performance.

In practical terms, the stresses imposed by the direct effects of high temperatures are more apparent (such as changing behaviour) and can be addressed more easily. However the indirect effects are less visible and may not be apparent until they become more serious with long-term consequences.

The severity of climate stress depends on many factors, including the actual temperature and humidity; the length of the heat-stress period; the degree of night cooling that occurs; ventilation and air flow; cow breed and size; level of milk production and dry matter intake prior to heat stress; housing type, overcrowding, aspect; water availability; coat colour, if exposed to sun; and hair coat depth.

How cattle lose body heat

The basic thermoregulatory strategy of cattle is to maintain a body temperature higher than ambient temperature to allow heat to flow from the animal via four basic routes of heat exchange, namely:

- conduction, in which heat moves from a warmer to a cooler surface. For this to occur, cows need direct contact with the surface;
- convection, in which the layer of air next to the skin is replaced by cooler air;
- radiation, in which heat can radiate from a warmer to a cooler environment; and
- evaporation, in which sweat or moisture is evaporated from the skin or respiratory tract.

The first three routes require a thermal gradient in which air temperature is lower than body temperature. Once air temperatures approach body temperature (around 39°C), the only viable method of heat loss is evaporation. This requires a vapour-pressure gradient to be effective, so is very dependent on the humidity, or amount of moisture in the air.

As cattle have a limited ability to sweat, the main route of heat loss in cattle during hot weather is evaporative cooling from the respiratory tract, namely the nasal passages and lungs. Cattle then increase their breathing rate to increase movement of air over the moistened surface of the upper respiratory tract and mouth. If humidity levels are high the effectiveness of this evaporative cooling is decreased and cattle may be unable to dissipate accumulated body heat.

Effects of heat stress

There are many adverse effects of heat stress, such as reduced appetite; reduced milk yields; reduced milk protein and fat levels; increased live weight loss; increased somatic cell counts hence mastitis; reduced fertility; and reduced calf birth weight.

Heat stress affects reproductive performance adversely in three ways.

1. Acute stress can lead to embryo reabsorption while chronic stress upsets normal cyclic status, through hormonal changes, particularly if cows are exposed to six hours or more to temperatures above 27°C.
2. In late pregnancy, reduced foetal growth can also result from heat stress, leading to increased calf mortalities.
3. The intensity of expression of oestrus is depressed, in that oestrus periods are shorter (for example 12 versus 17 hr) and although cows do cycle during hot periods, the percentage of those actually observed can be as low as 35–40%. This can be partly overcome by more frequent observations, such as every six hours rather than 12 hours.

Milking cows are maintained in a variety of environmental conditions. Without access to shade, the heat load on cattle grazing at pasture is generally lower than for cattle in yards, because the dirt surface absorbs less heat then grass, thus radiate more heat onto the stock. For example, the surface of a dirt yard can reach 60–80°C (on a day with high solar radiation and ambient temperatures of 40–45°C), but it will cool down rapidly once the sun sets. Clearly, access to shade, whether at pasture or in yards is highly desirable in regions with high radiation heat loads.

There are many symptoms of heat stress, with those ones more relevant to shedded cows shown in italics (Moran 2005). The initial signs are behavioural while the last five signs are the more severe physiological ones due to a failure to cope hence requiring immediate attention to reduce their adverse effects on cow performance. In order of increasing severity, they are:

- body aligned with direction of solar radiation;
- increased respiration rate;
- seeking shade;
- *refusal to lie down;*
- *reduced feed intake and/or eating smaller amounts more often;*
- crowding over water trough;

- body splashing;
- *agitation and restlessness;*
- *reduced or halted rumination;*
- grouping to seek shade from other animals;
- *open-mouthed and laboured breathing;*
- *excessive salivation;*
- *inability to move;*
- *collapse, convulsion, coma; and finally,*
- *physiological failure and death.*

The Temperature Humidity Index

The best single descriptor of heat stress currently available for dairy cattle is the Temperature Humidity Index (THI), as this combines temperature and relative humidity into a single comfort index. It does not account for solar radiation and air movement. The higher the index, the greater the discomfort, and this occurs at lower temperatures for higher humidities. The THI is presented Figure 9.2 while its effect on cow performance is summarised in Table 9.2.

For Friesians producing 20 kg/d, a THI above 78 leads to a decline in milk yield. There is also a decline in milk composition (milk fat and milk protein contents) but this occurs at 1–2°C higher than corresponding break points for milk yield. Very high milk yields, 35–45 kg/cow/d, can reduce the threshold temperature for heat stress by 5°C.

With regards reproduction, this declines before milk yield, namely at THI of 72. Cows in early pregnancy (up to three weeks) can abort while cows in mid pregnancy can have reduced birth weights. Cows are also more likely to have shortened and/or silent heats (less than eight hours). Heat stress delays oestrus

Table 9.2 Effects of Temperature Humidity Index (THI) on dairy cow performance.

THI	Stress	Comments
<72	None	–
72–75	Mild	Dairy cows adjust by seeking shade, increasing respiration rate and dilution of blood vessels. Reproduction is adversely affected. Milk yields will be reduced in high yielding cows.
75–78	Moderate	Both saliva production and respiration rates increase. Feed intakes decrease while water intakes increase. Milk production and reproduction are both reduced.
78–82	High	Cows become uncomfortable due to panting, high saliva drooling and high body temperatures. Milk production and reproduction will markedly decrease
>82	Severe	Potential cow deaths can occur

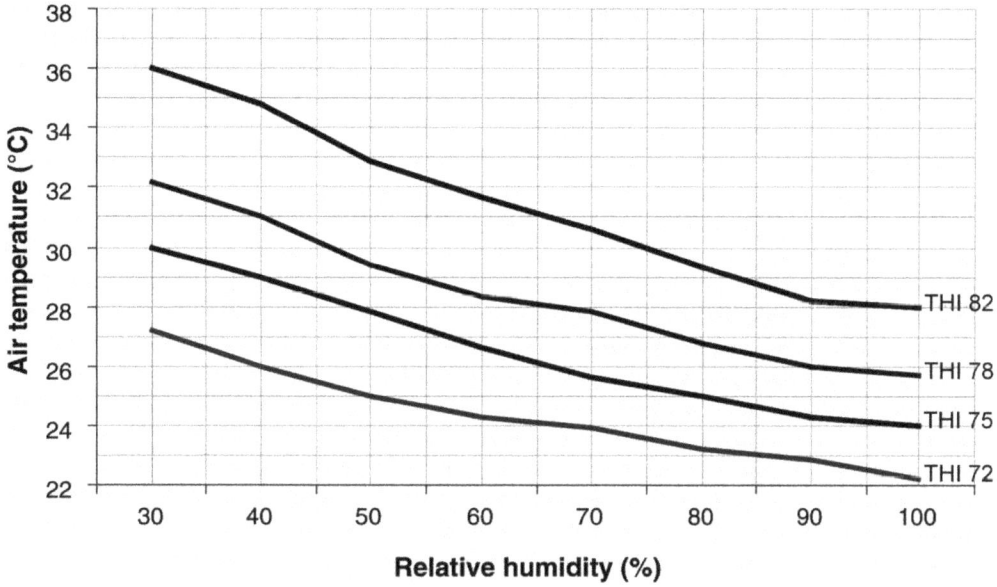

Figure 9.2   The Temperature Humidity Index for different combinations of temperature and humidity.

(hence submission rates) and, from five weeks prior to until one week following insemination, it can reduce conception rates and increase embryo mortality. By comparing conception rates between seasons (hot versus cool or wet versus dry), heat stress may be diagnosed as a problem if seasonal conception rates differ by more than 10–12%.

Cows are particularly vulnerable at temperatures above 30°C or, above 25°C with high humidity. Cows producing more than 15 L/d of milk are more susceptible to heat stress due to their higher metabolic heat load. Jersey cows are less susceptible than Friesians because of their higher density of sweat glands. When planning strategies to minimise heat stress, it is then important to give priority to non-pregnant cows, usually in early lactation.

Adverse effects of heat stress are delayed by several days. The effect of mean THI 2 days earlier has the greatest influence on milk yield, while the effect of mean temperature two days earlier has the greatest influence on feed intake.

Monitoring respiration rates

Observing the behaviour of cows is important in deciding when to modify management. If respiration rates reach 60 breaths/min, milk yield and reproduction may be compromised; this corresponds to 39°C body temperature, in contrast to a normal body temperature of 38.5°C. Higher yielding cows have faster respiration rates, because of the extra body heat production associated with higher

Figure 9.3 Checking the respiration rate for signs of heat stress.

feed intakes and milk yields. For such animals, if respiration rates exceed 80 breaths/min in 70% of the cows, it is indicative of heat stress. Certainly, when they exceed 100 breaths/min, cooling strategies should be introduced.

Respiration rates are easy to monitor (see Figure 9.3). Ensure the cow is standing or lying in a relaxed state and preferably cannot see any people. To improve accuracy, the recorder could move their hands in time with the flank movements until they are at a steady rate. Using a watch, they should count the abdominal movements for 10 sec, repeating the exercise until the count is consistent. Multiplying this by six will give the respiration rate in breaths per minute.

It is important to monitor stock of different ages and stages of lactation. Heifers are generally more resistant to heat stress because of smaller live weight and lower milk yields. High-yielding cows are less resistant because they have greater feed intakes hence internal body heat production.

Monitoring respiration rates at various times of the day is a useful tool in assessing the suitability of sheds for milking cows. If rate exceed say, 60 breaths/min in the morning, prior to the shed heating up, it is likely that the cows would benefit from simple modifications in their environmental management. It is unlikely that major modifications in shed design could be justified, such as increasing roof height or pitch or shed height at the side, although serious consideration should be given to constructing roof vents. If minor improvements cannot be made in the shed's natural ventilation, such as removing obstructions to the prevailing breeze, fans and/or sprinklers should be installed.

Clinical signs of heat stress

The following signs were developed for beef cattle, but they provide a guide to assess the degree of heat stress in milking cows:

- Mild heat stress – drooling, increased respiration to 80–100 breaths/min.
- Moderate heat stress – drooling, respiration of 100–120 breaths/min and occasional open-mouth panting.
- Severe heat stress – drooling, respiration rate greater than 120 breath/min and open-mouth panting with tongue out. Cattle also have an agitated appearance, hunched stance and will often have their head down.

Cattle can move from mild to severe heat stress very quickly, within 30 min to a few hours. Therefore extra vigilance is required once mild heat stress is detected.

Observing the behaviour of cows is important in deciding when to modify management. Improvements in milk yields of up to 3–5 kg/d are possible through effective cooling strategies.

Feeding management during heat stress

The digestion of fibre produces more heat in the rumen than the digestion of other carbohydrates. Therefore, offering much of the TMR during the cooler periods of the evening will reduce internal heat production, particularly if it is high in fibre. Rations high in rumen degradable protein can also depress appetite during periods of heat stress. Cows consume about two-thirds of their feed during the cooler evening and night. Feeding smaller amounts more frequently will reduce the likelihood of forages becoming dried out and losing their palatability.

The temperature of the drinking water has little effect on water intake or heat balance. Feed the best quality forages at night and feed more concentrates during hot periods. Offer more salt to replace the minerals lost in sweat. Cows lose a lot of potassium in their sweat, unlike humans and horses whose sweat is high in sodium concentrations.

Heat stress can lead to higher incidences of lactic acidosis. Depressed feed intakes will firstly reduce saliva production, which buffers the rumen against rapid changes in pH, and secondly, reduce rumen contractions hence movement of digesta out of the rumen. Furthermore, rapid respiration rates for lengthy periods can reduce the concentration of sodium bicarbonate in the saliva, reducing its buffering capacity even further. In addition, when offered the opportunity, cows may prefer to eat concentrates and reject forages, predisposing them to acidosis.

Fats contain over twice as much energy for the same quantity of cereal grain, so are potentially valuable energy supplements when feed intakes are limited, as in hot weather. Adding fat does not necessarily alter feed intakes but can improve milk yields. If too much is fed, fibre digestion will suffer. A number of sources are available, such as vegetable oil and commercial 'bypass' fat supplements. The high-yielding cows are more likely to benefit from these but it is best to consult a nutrition adviser before embarking on such a feeding plan.

Reduced forage intakes can decrease milk fat contents, while milk protein contents may fall due to lower dietary energy intakes. The immune system of heat-stressed cows being under greater strain, would make them less able to cope with subclinical mastitis, which would show up as higher levels of somatic cells in the milk. Not only would heat stress reduce milk yield, but its lower milk solids and higher somatic cell counts would reduce milk returns even further through lower unit returns for the milk.

Provide pasture at night when conditions are cooler and cows are more likely to want to graze. On hot days, cows should be milked and fed well before 1000 hr. On days with predicted heat waves, it should be done by 0900 hr. Once they have eaten, grazing cows should be moved to the coolest area on the farm, such as under trees or other shade, but with the opportunity to have another feed if they so desire. Providing them with sprinklers allows for more efficient cooling so long as there is a breeze, either a natural one or using cooling fans. Avoid walking cows when it is really hot and consider milking later in the afternoon than normal.

Developing a heat stress management program

With climate change increasing the likelihood of lengthy periods of heat stress on Australian dairy farms, it is important to develop a farm program to cope with heat stress using a variety of strategies. Managing cows to produce more milk increases their susceptibility to heat stress hence their need for such a program. Dairy Australia (2008c) has listed some of the key strategies for grazing cows. These can be modified for cows on feedpads, as follows:

- With solar radiation the major heat load, the first approach should be to protect stock from direct sunlight through provision of shade. Shade cloth is not expensive and all feedpad systems should at least have some structure where such material can be installed over the hot summer months. Not only will it reduce heat load hence improve cow comfort and performance, it will minimise the possibilities of animal welfare concerns once temperatures exceed 30°C. Obviously permanent and more effective shade structures are more desirable. Trees may be effective in paddocks, but they would rarely provide sufficient shade in a feedpad because of the stocking density.

- At high temperatures, cutaneous evaporation is a primary means of dissipating heat in cattle. Dairy Australia (2008c) recommend sprinklers be fitted into every dairy yard in Australia. For grazing herds, this would be ideal, but cows spend longer periods of time on feedpads, so sprinklers in feedpad system would be a better investment.

- In humid areas, fans are necessary to maximise the benefits of sprinklers. The summers in southern Australia usually do not coincide with humid weather. Although supplementary air movement will improve cutaneous evaporation rates, the cost-benefit ratio of fans in the southern dairy regions needs to be evaluated more fully. In northern Australia, like any area that experiences potentially humid summers, fans would be a good insurance to complement sprinkler systems.

- Changing the times for milking is an easy and effective management strategy, even for cows on feedpads that do not have to walk long distances. Milking earlier in the morning and later in the evening will reduce heat stresses for both cows and staff.

- Better access to cool drinking water should be easier on a feedpad than at pasture because the water is less likely to heat up in the pipes over the shorter distance. Shortages of drinking water will be more apparent, hence can be corrected, on a feedpad where feed intakes are more easily monitored than at pasture. Farmers with feedpads would already have considered the adequacy of water supplies in their initial planning and construction.

- Summer nutrition programs may be less of an issue with feedpad farmers than those entirely pasture based. Presumably feed intakes are already closely monitored and with PMRs, fibre intakes would be under more control, thus acidosis problems should be minimised. Sub-clinical acidosis can be a major problem in grazing cows due to the slug feeding of concentrates at milking and changes in grazing behaviour in response to heat stress. Ensuring the high yielding cows are fed good quality forages while replacing some of it with extra grain may prove beneficial during hot summers. Increasing levels of minerals (sodium, potassium and magnesium), changing to slower fermenting starch sources (such as maize or sorghum) and feeding more rumen buffers are good feeding practices to consider during summer.

- If cows are still grazing each day, evening would be a more appropriate time to offer them the pasture. On very hot nights, this can even provide a better environment to cool off, because of lower ground temperatures, more space to move around and greater access to air movement than on a feedpad.

- Mating management may require some fine-tuning as heat stress can severely affect conception rates, particularly in high-yielding cows.

Dairy Australia's Cool Cow program offers a website with information for farmers on dealing with heat stress, including weather alerts during summer for their particular region (see http://www.coolcows.com.au/tools/weather-forecaster).

Designing feedpads to minimise heat stress

Roofing

Assuming the sides of the shed are open, to allow maximum ventilation, greater use can often be made of the principles of air movement when designing the roof. Because hot air rises, and considerable heat is produced by a concentration of milking cows in a shed, there should be an opening along the top of the roof, with a cap over it to restrict rain entering the shed. The roof slope should be greater than for feed sheds, namely 3–4° per 2.3 m, with the opening at least 50 cm wide, and the full length of the shed. A second design is a roof slope of 33° (4 in 12), with a vent at the top of 30 cm plus 50 mm per 3 m of width for sheds with more than 6 m wide. The lowest point of the roof should 3 m from the ground. The steeper roof pitch increases airflow across and above the roof, thus creating negative pressure over the opening. This hastens the flow of air out the top as well as creating turbulence of air movement around the cows. If farmers are still concerned about rain entering the shed through the vent, a gutter system can be installed below the ridge opening while the concrete floor can be sloped away from the feeding area.

The roofing material should be aluminium, white or galvanised iron. Shade cloth that blocks 80% of the light can provide cheaper protection but should not interfere with ventilation within the shed. The lowest point of the roof should be at least 3.7 m from the ground. Insulation under the roof is not necessary while spraying water onto the roof may only be effective in reducing roof and shed temperatures in areas with low humidity. To create air movement through the shed, the roof should slope at 19° and be vented at the top, allowing a width of 300 mm + 50 mm per 3 m width of the shed when greater than 6 m. Each cows should be provided with an area of 3.5–4 m^2 (including troughs) when housed for up to 12 h/d.

Flooring

Mastitis will be minimised with concrete floors and an adequate waste handling system. The floor should have a 2–3% slope to allow for a flood washing system with nib walls 200 mm high on the side of the flush bays. A slope greater than 3% can increase the slipping of tractor wheels when pulling feed wagons while a slope of less than 2% can still be flood washed when cleaned in combination with mechanical scraping. Routine mounding of fresh sand or compacted gravel under the shed can help if the shed does not have a concrete floor and is excessively wet.

Orientation

The shed should be sited so that wind breezes are not blocked by any obstacles or other buildings. Ideally it should be on the highest ground possible, which will also be good for drainage of effluent, with other buildings located downwind on the site. There should be a minimum of four times the height of the nearest wind barrier as a horizontal separation. The ideal orientation, from a ventilation point of view, would allow the prevailing winds to hit the shed perpendicular to the side. This allows the wind to travel the shortest distance before exiting the shed, to improve the rate of air exchange and provide the cows with fresh air. The longer the shed, the more important is this perpendicular orientation to the prevailing winds. Other factors to consider are exposure of the outside stalls to sunshine, future expansion plans, cow flow, traffic flow and manure flow.

A north–south orientation is preferable over compacted clay or gravel feedpads, to allow the sun to dry underneath both sides of the shed. An east–west orientation is recommended over concrete feedpads, where the southern side can dry out properly so as not to predispose the cows to mastitis. Trees should be planted on the western side of the shed to reduce solar radiation. Eaves that extend one-third the side height will provide good sun protection.

The water should be located on the eastern and western edge of a north–south shed to reduce walking to water. Feed toughs should be positioned under the shade. The feed and water should be located on the southern and northern sides in an east–west shed.

Installing sprinklers and fans

Sprinklers should be suspended 2.3 m from the ground with water directed onto the back of cows. A filter should be installed at the beginning of the waterline while sprinkler nozzles should be easily removed for cleaning. These nozzles (90–360^{0}) should be directional so they can be adjusted to reduce wetting of feed during any major shifts in the prevailing wind.

In humid climates with little breeze, both fans and sprinklers are required to cool cows evaporatively (see Figure 9.4). The rate of airflow will determine the efficiency of sprinklers in cooling cows in low humidity environments. An ample water supply is obviously required, namely 100–150 L/cow/d. The cows should be under cover and provided with concrete floors (with a 2–3% slope) where sprinklers are installed. In high humidity environments, sprinklers that spray out large water droplets to wet the cows' skin are more effective for cooling. In hot, dry areas, a fine water mist can cool down the air temperature.

Figure 9.4 Fans and sprinklers are essential for climatic control in hot and humid areas.

Cows should only be sprinkled for 1–3 min every 15 min, applying 1–2 mm water. Feed and water should be in close proximity to the cooling area. The system should be run on a remote control valve (solenoid) sprinkler system with a 15 min cycle which shuts down at temperatures below 26°C. The pipe size will depend on the length and area of the facility, the number of sprinklers and their low rates, ranging from 32 mm diameter for up to 30 m length, 51 mm diameter for up to 60 m length and 76 mm diameter for up to 150 m length, with smaller pipes within this 150 m long facility. Low-pressure sprinklers and regulators work best (that is 0.70 kg/cm^2) as lower velocity gives less mist and less spray drift.

The system should be installed next to the feeding area with 180° nozzles in a wide feeding area and 360° nozzles in the centre of the cow alley under the fans. The nozzles should be spaced out twice the radius of their throws.

Fans, on the other hand, can be installed in many ways. The direction of the flow should be with the prevailing wind and the size of the fans and their direction depend on spacing of posts in the shed. A 0.5 hp, 0.91 m diameter fan rated at 5–6 m^3/min will blow a distance of 9 m while a 1.0 hp fan, 1.21 m in diameter and rated 9–10 m^3/min will blow a distance of 12 m. In wide feeding sheds, the side-by-side spacing width of 0.9 m fans should be about 6 m, whereas 1.2 m diameter fans can be spaced every 9 m. Fans should be tilted to blow down on the floor directly under the next fan, which is about 30° from vertical. Further details of feedpad designs and facilities to reduce heat stress are provided by Dairy Australia (2008c).

Minimising cold stress in dairy herds

Although Australia's dairy regions rarely suffer from severe winter weather, there have been instances of extensive losses of dairy stock through unpredicted cold stress, sufficient to justify its mention in a book on feedpads. Unseasonal cold weather can have drastic effects in stock not used to this extreme. During the middle of a hot summer (in February 2005) in northern Victoria, the air temperature fell from 38 to 13°C over a 24 hour period, together with intense wind and heavy rain during the night. Over 2000 dairy cows and young stock in the region died during that night from cold stress, and also from trauma as they sought shelter around small windbreaks. All-weather feedpads, with protection against winter as well as summer climatic extremes, could have saved many of these animals.

The effects of low air temperatures, wind, wet hair coat and mud are additive. Growing cattle can be comfortable within a wide range of temperatures, from –5°C to 20°C, depending largely on the length and condition of the hair coat (dry versus wet, clean versus muddy).

The lower critical temperature (LCT) is the temperature at which energy intake must increase to minimise reduction in weight loss in growing cattle or prevent weight loss in mature cattle. In beef cattle, this can vary from 15°C for stock with a wet summer coat, to 0°C for a dry winter coat to –8°C for a dry heavy winter coat.

The combined effects of temperate and wind are often expressed as the wind chill index and this is used when considering the severity of cold stress. For example, at 2°C with no wind, the wind chill index decreases to –2°C with 8 kph wind, to –10°C with 24 kph wind and to –13°C with 32 kph wind. Therefore it is important to reduce wind exposure at low temperatures to reduce cold stress in cattle.

In general, the energy requirements of a dry non-lactating cow increase by 2% for each 1°C below 0°C whereas if she is wet, the increased energy requirements begins at 15°C and increases 3% for each 1°C drop.

The LCT also varies with the live weight, body condition and energy intake of cattle. These vary from:

- –30°C to –40°C for dairy cows and feedlot cattle on high grain diets because of their high-energy intake and metabolic rate.
- –25°C for cows in good body condition.

- −17°C for thin cows, indicating that thin cows need preferential feeding management.
- −10°C and −25°C for non-lactating dairy cows and growing calves, in good body condition.
- 1°C for dairy calves from three-weeks old to weaning.
- 13°C for young calves, from birth to three weeks of age.

The effect of acute cold stress is reduced with acclimatisation. This can last for the entire winter. When the temperature drops below the comfort (or thermo-neutral) zone (15°C to −20°C), certain adaptive changes occur. Hair coat becomes thicker and longer and metabolic rate increases. Feed intakes can increase by 30%. This increases energy requirements by 10% for every 10°C in effective temperature below the comfort one. Cows in good condition can increase their hay consumption by 5 kg, but to provide more energy, this must be supplied as cereal grain or better quality forage.

Because milking dairy cow are typically housed in regions with severe winters, and fed a high-grain diet, they are relatively immune to the effects of cold stress. Milk yields can decrease, however, if the cows are already consuming their limit for feed and the energy density of their diet cannot be increased.

In a well-insulated shed, the milking herd requires little additional heat other than that produced by the animals until the outside temperature falls below −4°C. Calf and maternity areas may require supplementary heat as will the milking shed to keep it dry and above freezing. The recommended range of temperatures (and humidity) for different classes of stock are:

- Milking cows: −7°C to 24°C and 25–75% humidity;
- Calves (under 6 weeks): 10°C to 27°C and 25–75% humidity; and
- Other young stock: −18°C to 27°C (if draught-free).

The effects of severe cold stress

Hypothermia occurs when the body temperature drops well below normal (38°C). Mild hypothermia occurs at body temperatures from 30°C to 32°C, it is moderate at 21°C to 29°C and severe below 20°C. Once it drops below 28°C, cows cannot return to normal without assistance through warming and administration of warm fluids. As hypothermia progresses, metabolic and physiological processes slow down, the blood is diverted from the extremities to protect the vital organs and teats, ears and testes are prone to frostbite. In extremes, respiration and heart rate drop, animals lose consciousness and die.

In most situations, when cows are subjected to temperatures below their LCT, but without signs of hypothermia; this increases their maintenance requirements as they adjust to the conditions and divert more energy to maintaining body temperature. There are two potential responses to this situation:

1. Cows have access to higher quality or more feed so can maintain their body weight. Compared to –1°C, when exposed to –12°C, cows require 20% extra energy, which can be provided by 1.0 kg extra grain or 1.7 kg extra hay.
2. Cows don't have access to increased feed and lose weight. Cows must 'burn' body mass, usually body fat, just to maintain body temperature. They start to enter a downward spiral because the more weight they lose, the less body insulation hence they lose weight even faster.

Cows and especially heifers that lose weight, calve down in poor condition. This leads to increased calving difficulties, more lighter and weaker calves born and a higher calf mortality. These dams produce a reduced amount of colostrum, of lower quality. In addition, they have lower milk production, increased calf mortality and reduced growth rate in surviving calves. These cows usually have delayed return to oestrus, longer days open and poorer reproductive success.

Management strategies to combat cold weather

Although farmers cannot control the weather, they can do many things to relieve the effects of cold stress to reduce costs and improve production efficiency. These include monitoring the weather and providing extra feed in response to cold weather, particularly to cows in their last three months of pregnancy; constructing simple windbreaks or shelters so cows can cope better with extreme temperatures. Forcing cattle into sheds during storms for long periods, however, can increase their chance of disease or getting wet, due to condensation dripping off the roof, the longer they remain in close quarters. Pasture offers the advantage of space, fresh air and confident footing; providing adequate dry bedding to keep cows clean and dry as wet and/or muddy coats reduces their insulation properties greatly; providing adequate drinkable water to ensure cows will consume more of the feed on offer. Frozen troughs and excessively cold water will seriously limit water intake; splitting the herd into small groups, to allow thin cows to access better quality and/or more feed. Competition between cows often leads to the timid, smaller or younger cattle not receiving their fair share; feeding cattle late in the afternoon or early evening so that the energy from that feed is more effective in warming the stock. Incremental heat production is at its maximum 4–6 hr after feed is consumed. Consider providing specific facilities for rewarming sick young calves, such as heat lamps or warm blankets, ensuring that attention is given to ventilation in and sanitation following use of such a warming box. Calf coats can improve insulation of sick calves while facilities to heat milk or drinking water can also provide additional methods to warm sick calves.

10

Farm management

This chapter covers issues for farm management to consider when training and familiarising staff with feedpad technology and machinery for mixing and feeding out Partial Mixed Rations.

The main points in this chapter:

- Farm staff should see any changes to their job routines as different ways to work and not just more work.
- Occupational Health and Safety should be top priority on any farm and the owner of the farm must provide a safe working environment for all staff.
- As well as increasing their surveillance for new animal health and welfare problems, stock handlers require a good understanding of the behaviour of confined cattle.
- There are many potential hazards in maintaining an effective effluent system which staff need to be aware of and take necessary precautions.
- To avoid or minimise the risk of adverse impacts on staff safety and animal welfare, Standard Operating Procedures should be developed and implemented for all aspects of farm operations and each piece of machinery. This includes developing contingency plans for all foreseeable emergencies.
- Machinery for feedpad operation can constitute a major investment, so such purchases should be carefully planned. Such equipment needs to be sufficient for future herd requirements.

Feedpad technology generally implies that the dairy farm owner or manager plans to increase the farm's milk output through intensifying feeding practices. As well as investing in new equipment and feeding facilities, farm staff will require additional training in feeding and managing confined dairy stock. This will include both their contribution to the feeding program, such as preparing and

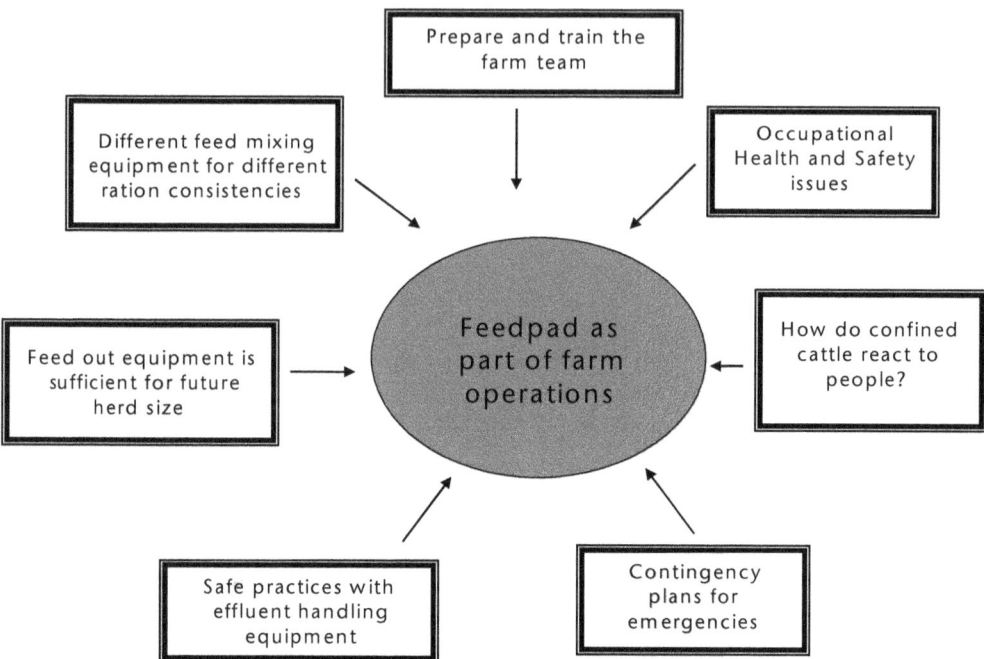

Figure 10.1 Some of the key considerations for good farm management.

feeding out the Partial Mixed Ration (PMR), as well as to herd management; for example, increasing surveillance for new animal health and welfare issues.

This chapter covers issues for farm management to consider when training and familiarising staff with the new technology (see Figure 10.1). Requirements for machinery for mixing and feeding out rations on the feedpad are also included. Machinery for handling and recycling feedpad effluent, however, are not discussed as they have recently been reviewed extensively in the Dairy Australian (2008b) effluent and manure management database.

Preparing the farm team

Introducing feedpad technology onto a farm creates new tasks, different work routines and potential safety hazards compared to pasture-based feeding. In addition to managing the finances and cows, farmers need to plan ahead to ensure their farm team has the skills for these new tasks.

When selecting the feedpad development program, it is important to consider who operates it on a day-to-day basis. Compared to pasture-based systems, less time will be taken up by activities such as irrigating and fertilising pastures and moving stock around the farm. The workload then becomes more focused on delivering feed to the cows. For example, it is important to have the formulated ration delivered to the feeding area while the cows are being milked so it will be on offer at that time of the day when they are likely to be their most hungry. It will also keep them standing during the 30 min post-milking period prior to their teat canals being sealed, the time when they are most susceptible to invasion by the environmental bacteria causing clinical mastitis.

Farm staff should see the changes as just different ways of doing their job and not just more work. Following the establishment of the feedpad system, jobs and responsibilities should be delegated to farm staff who have been trained, and have shown competence, in the new skills and procedures. Farm managers should be aware that in addition to developing technical skills and knowledge, there are other personal characteristics affecting staff ability to become competent in new tasks. These include job motivation and commitment, job satisfaction and personality.

Rosters should be organised to minimise unnecessary work during weekends and after normal work hours. Fitting jobs in with milking routines can reduce overtime.

Apart from actually mixing and delivering the feed, the feeding system should be monitored regularly for its effectiveness and efficiency and to keep track of feed reserves. Keeping written records of amounts of all feed ingredients delivered by suppliers and all feeds mixed and fed to stock is important to avoid feed shortages or unaccounted bills to be paid. Milk responses should also be closely recorded to modify ration formulations, if necessary.

Feedpad technology requires very careful observation skills to achieve high milk yields hence high farm profits. For example, routine observations on feeding behaviour, rumination and manure consistency can help fine-tune ration formulations. Signs of excessive feed wastage require changes in either methods in feeding out or the amounts offered. Staff can identify animals that are sick, weak or injured and can carry out the appropriate actions. Signs of heat stress are easy to quantify. Respiration rates can be quickly monitored by counting flank movements for 10 seconds then multiplying by six. If cows breathe faster than 60 breaths/minute, heat stress may be affecting feed intakes adversely. Cows on heat may prefer to be mounted on earthen floors rather than more slippery concrete floors. Some cow behaviour reactions appear to make little sense in human terms. For example, cows may all crowd together under one shade on a hot day while another nearby shade is empty. Are staff numeracy skills acceptable when measuring out feed ingredients? Have any of the staff developed the skills to monitor and interpret any changes in milk composition? Perhaps interested farm staff may be motivated sufficiently to want to learn some new nutritional skills so they can better understand the new feeding systems being implemented. Are there up-to-date 'Standard Operating Procedures (SOP)' on all relevant farm equipment and does the farm team all fully understand them? The legal aspects of animal welfare have been covered in Chapter 9. Perhaps some of the staff may have computer skills that

Figure 10.2 Linear module feeders can also be used to feed out concentrates.

could be utilised for data collation and manipulation using software such as MIFC10 (see Chapter 6).

Occupational health and safety issues

Occupational Health and Safety (OH & S) must be a top priority on any farm since the manager/owner is ultimately responsible for its implementation. The owner of the farm must provide a safe working environment for all staff, contractors and visitors which includes if necessary providing appropriate training. Of more importance to the owner, he must address the regulatory obligations of relevant government legislation. These are discussed in Chapter 12.

Relevant questions to ask include whether the staff are using safe equipment that is suitable for the purpose? Can all staff identify the hazards and minimise the risks associated with different machinery? For example, ensuring all guards and safety systems are in place on all machinery and equipment. Is there sufficient lighting around high use areas, especially those likely to be frequented before sunrise or after sundown. Can staff rectify failures of equipment or machinery promptly and put in place repair and maintenance programs? You must maintain or replace all personal protective equipment regularly. Children must be closely supervised in all areas of feedpad systems. Staff should be vaccinated for Q fever, as should livestock for leptospirosis. With increasing amounts of conserved forages being purchased, will their long-term storage require extra OH & S responsibilities,

Figure 10.3 In case of hay fires, machinery should not be stored in a partly filled hayshed.

such as covering large silage bunkers or being more conscious of hay fires in large haysheds or stacks (see Figure 10.3)? Do staff consider topography and areas around ponds when using slurry tankers and solid spreaders, which are not well suited for any steep, sloping land. Are they all aware of the risks in working in feed storage and mixing areas, such as unloading trucks, safety around haystack and silage bunkers and feed commodity sheds?

People, machinery and cows all share the same space on the feeding area. Is there a system in place to ensure everyone's safety? Moving large numbers of stock to, from and around concrete feedpads that are likely to be quite slippery at times, will introduce many new hazards to unskilled staff. Milk production is generally higher from calm cows, and confinement feeding has introduced a new series of stress to the milking herd. How are the cows coping and how have the staff adapted to it? Staff may have developed some of the skills of stock handling while moving cows to and from the paddocks or in the milking shed, but can they anticipate their behaviour when confined to a feedpad? The next section discusses this in more detail. Does the team know what to do if something goes wrong (see later section on contingency planning)?

With regard to the important issue of human health, staff should wear appropriate protective equipment for activities involving direct contact with manure, chemicals and veterinary products. They should be up-to-date with relevant inoculations, such as Q fever and leptospirosis. They should avoid unnecessary handling of manure, minimise spray drift when recycling effluent, and avoid having to use recycled effluent inside milking sheds or vat rooms.

Understanding how confined cattle react to people

It is essential that farm staff become familiar with the way confined stock react to people. How much do they really know about the basic sight and hearing senses of cattle?

Firstly, cattle have wide-angle vision, if fact they can see 300° out of 360° around them. They use this field of vision to define their 'personal space', which we call their 'flight zone'. Secondly, cattle are quite sensitive to high frequency noises, and compared to people (who can hear noises from 1000–3000 hertz), they can hear noises up to 8000 hertz. Further general principles of stock behaviour include:

1. When a person moves into their flight zone, cattle will normally try to move away.
2. The size of their flight zone will decrease slowly if they are handled frequently and gently.
3. Previous experiences will affect how animals react to future handling, with memories persisting for many months. Obviously fear memories are significant in increasing flight zones.
4. Cattle can tell the difference between two situations readily and make choices to avoid the more stressful one.
5. Cattle are sensitive to changes in colour and texture.
6. Moving objects and people seen through sides of a chute can frighten animals.
7. Novelty is a strong stressor, while repeated exposure will reduce the novelty effect.
8. Cattle are herd animals and don't like to be separated from their herd mates.
9. Groups of cattle that have body contact remain calmer.
10. Unexpected loud or novel noises can be highly stressful.
11. Cattle adapt to reasonable levels of continuous sound such as background noises or music readily.
12. Cattle exposed to a variety of sounds, such as radios with talk and music, may have a reduced reaction to sudden noises.
13. Cattle adapt to handling readily, even if the events may be initially stressful, such as walking up a race, into a head bale or being transported.
14. Cattle can be trained to accept restraint voluntarily with relatively low levels of stress.
15. A small amount of inconsistency in care and handling can reduce their stress response to new sights and sounds.
16. Consistent poor handling can create chronic stress.

Safety issues with effluent systems

As well as specific OH & S issues for staff handling stock and operating machinery, there are various hazards in maintaining effective effluent systems. Storage, pumping, spreading and cleaning feedpad effluent systems can release large amounts of gases from decomposing manure. Hydrogen sulphide (highly toxic), carbon dioxide, ammonia and methane are all dangerous for farm staff. Where feedpad wash is piped directly to the effluent pond, a water seal or gas trap could be installed to prevent such gases from entering confined spaces. Effluent ponds can form a substantial crust that supports weed growth and looks like solid ground. Ponds must be fenced off with warning signs making it obvious of the dangers. Place a rescue rope and float within the fenced-off area around the pond. Tractors should not be used near the edge of effluent ponds. If there is no alternative, low barriers to the pond or chocks to prevent the tractor from moving backwards should be installed. Flood-wash tanks must be installed on stable foundations and supports. Sumps and solid traps must be covered or fenced off to exclude children and stock. Manure pits must not be entered without a respirator and an emergency plan. Exposed moving parts of effluent pumps must be guarded. Control excessive weeds and vegetation around the manure management system (especially ponds) so the extent of the system is clearly visible. Place wheel-chocks or railing barriers with pond-agitation machinery to prevent vehicle movement. Maintenance and desludging operations require extreme caution as clay surfaces can become slippery when wet.

Contingency plans for emergencies

Standard operating procedures (SOP) should be developed and implemented for all aspects of farm operations and each piece of farm machinery, to avoid or minimise the risks of adverse impacts on animal welfare, staff safety and amenity and the surrounding environment. Contingency plans, however, are necessary when responding to or managing accidents and other foreseeable emergencies. They should identify the trigger points for their implementation, along with target response times for critical incidents. Contingency plans should be considered for power failure (electricity, gas); water and feed supply failures; chemical or fuel spills; large feed spillages; drain blockages; manure management system breakdown; shed malfunctions, including milking parlour; extreme weather events (heat and cold stresses); natural disasters such as drought, flood and fire; odour or dust events; emergency disease outbreaks and catastrophic mortalities; higher than average stock mortalities; and the potential for inability to sell, hence transport milk off-farm.

A procedure manual should be developed to recognise such potential emergency situations and to provide clear strategies and measures to minimise those risks and potential impacts. It should also list the relevant statutory authorities, provide procedures to respond to complainants and how to investigate causes following a major environmental incident. Two of the key emergencies that require rapid action are the discharging of manure or recycled effluent into waterways and the mass mortality of stock.

Machinery for mixing and feed out

Machinery can constitute a significant part of feedpad investment, so purchases should be carefully planned. It is important to plan ahead and purchase tractors and feed out equipment sufficient for future herd requirements.

When moving to a feed out system using a mixer wagon, it may be necessary to purchase an additional tractor and front-end loader dedicated to the feeding facility. The existing tractor and machinery, such as a moderate-sized tractor and machinery for feeding out round or square bales in the paddock, should be kept as a back up. When purchasing a mixer wagon, it is important to consider the power requirements of the tractor needed to operate it. Different mixer wagons have different power requirements. For example, paddle mixers have lower power requirements than vertical or horizontal mixers. The tractor needs to be large enough to safely tow a full wagon on slippery or sloping surfaces. In their survey of dairy farmers in Queensland, Busby et al. (2007) noted that most farmers would have purchased larger mixing machines and tractors if they were starting their feedlot development program again. Table 10.1 presents indicative tractor power requirements for different sized mixer wagons (Amaral-Phillips et al. 2002).

Machinery then should be easy and safe to operate. It should fit in with existing farm infrastructure. It should deliver the same diet to all stock in the group, deliver a diet that adds farm profit and be easy to maintain.

When mixing and delivering feed it is important to check that staff follow the machine manufacturer's standard operating procedures for loading and mixing; that they know what a well-mixed ration looks like. Under-processing (with long fibre not well mixed) or over-processing (with no remaining long fibres due to too much mixing) can be a problem with inexperienced staff. Check if the feed trough is higher than 30 cm above the feedpad surface; if there enough trough space for all

Table 10.1 Approximate power requirements for tractors needed for mixer wagons of different capacities.

Effective mixer capacity (m³)	Tractor size* (HP)
<6	75
6–8	100
9–14	125
15–20	150

* Values are approximates only and should be checked with the recommendations from equipment manufacturers, as they may vary from these estimates.
(Source: Amaral-Phillips et al. 2002)

stock in the group; if there is sufficient access to water; if there have been any sudden changes of ration composition; if long fibre components in the ration are separating out from smaller sized feed ingredients during feed out or while the cows are eating; and if the ration is palatable. Is it too dry or too wet? Is it spoiling on the feedpad too quickly?

Key factors when mixing rations include the order in which feed ingredients are loaded; mixing time and speed (revolutions per minute); and dry matter content, hence addition of liquids. Follow the recommendations provided by the manufacturer.

Mixer wagons

There are three basic forms of equipment that can be used for mixing and delivering PMRs to stock on feedpads. Most farms have front-end loader buckets, which can be used for very simple mixing of dry rations. Silage carts are designed purely to deliver chopped silage but can also be used for crude mixing by layering the various feed ingredients in the wagon. Mixer wagons are purpose built for mixing a diversity of ration ingredients. Features of these various machines are presented below:

1. Tractor and front-end loader bucket
   - Can only mix dry processed feeds.
   - Time-consuming to mix and feed out.
   - Cannot mix minerals, feed additives or urea into rations.
   - Long fibrous material separates out.
   - Must be fitted with load cells to weigh ingredients.
   - Uneven mixing hence inconsistent stock feed intakes and performance.
2. Silage cart
   - Cannot incorporate long and chopped materials.
   - Cannot accurately mix minerals, feed additives or urea into rations.
   - Long material separates and tends to be wasted as cows can sort out less palatable ingredients.
   - Must be fitted with load cells to weigh ingredients.
   - Can have high wastage if cart has a wide feed-out.
   - Uneven mixing hence inconsistent stock feed intakes and performance.
3. Mixer wagon
   There are more than 20 different brands of mixer wagons available in Australia. Mixer wagons come with three different mechanisms, namely paddle mixing and vertical or horizontal mixing. All are capable of accurately weighing feed ingredients, evenly mixing feeds and delivering them to stock with minimum wastage. The volume capacity of these wagons depends on feed bulk density. All wagons provide good access to feedpads, but the wheel

Table 10.2 Features of different types of mixer wagons.

Features	Paddle mixer	Vertical mixer	Horizontal mixer
Mixing ability	Cuts forages for processing	Throws feed into air	Mixes feed horizontally
Weighing ability	Standard	Standard	Standard
Loading height	High	Low	Low
Water tight	Usually	Not always	Usually
Horse power requirements	Low	High	High
Paddock access	Good	Some brands do have good access	Some brands do not have good access
Other features	Relatively heavy No dead spots in mixer Slow moving so hard to over-process mix Simple drive with no gear box	Can be single or double cone Easily process dry ingredients Take care not to over-process mix Good capacity	Can have single, twin or more augers Some brands have dead pots in mixer Some will process large hay/silage bales Can compact feed

(Source: Dairy Australia 2007b)

arrangements and centre of gravity in some is not ideal for access to uneven paddocks. Details of each basic type are presented in Table 10.2.

When choosing a mixer, decide if it should be stationary or mobile. If mobile, is it for use in paddocks as well as on the feedpad? Will it fit easily through all feed passages and gates into paddocks? Will it feed into all feed troughs? What is the desired capacity? What is the likely largest group size to be fed on one ration? Can existing tractors handle the high workload requirements? Should it handle long-stemmed hay? Do you want a magnet to remove hardware from the ration at feed out? How sophisticated do you want the weighing system to be? Feed scales can be linked to computers if required. Should it be watertight? Do the suppliers provide good warranty and after sale service? Does the brand have a good maintenance schedule? Have the suppliers got ready access to spare parts? What options are there to purchase a new machine, including sourcing a second-hand one or even leasing a machine? Does the local supplier have the best deal for you? Some companies can design a diet feeder specifically for your requirements. The final price per cubic metre of wagon capacity is a good indicator of value for money. How long will it take for the mixer wagon to 'pay for itself'?

Computer programs are available to evaluate the economics of purchasing a mixer wagon (Amaral-Phillips *et al.* 2002) and to analyse whether the financial

Figure 10.4 The author feeding out hay using a linear module feeder.

benefits of reduced feed wastage and improved cow performance are sufficient to justify the extra capital investment of purchasing feed out equipment for a feedpad operation (Leddin and Armstrong 2009). The second program will be more fully discussed in Chapter 11.

11

Economics of feedpad technology

This chapter introduces the concepts of milk income less feed costs (or feeding profit) and partial budgets in addition to software to assist with investments in feedpad technology.

The main points in this chapter:

- The profitability of any dairy feeding system is largely driven by the cost of feed, the returns for milk and the relativity of the two. Milk income less purchased feed costs is a simple and useful measure of feeding profit.
- Computer software is available to calculate such profitability measures as milk income less purchased feed costs. A program is available free of charge from the author of this manual.
- Despite higher feed costs, there can be increased profits made with higher milk yields. This does not take into account the other variable and all the overhead costs associated with better feeding management.
- There are high costs of feed wastage associated with current inefficient feed out systems. Losses can vary from as high as 30% to less than 5% as feedpad technology improves.
- Such often-readily-accepted losses mean that the investment payback period from investing in more efficient feedpad systems can be less than two years.
- Computer software has recently become available to assist with making investment decisions on feedpad technology.
- The value of feedpad effluent is often underrated and can amount to more than $80/d for a 250-cow herd spending half their time on a feedpad.
- Partial budgets and sensitivity analyses are tools that should be used to assist in making the decision, 'Is a feedpad system a good investment for my dairy enterprise now, in the future or not at all?'

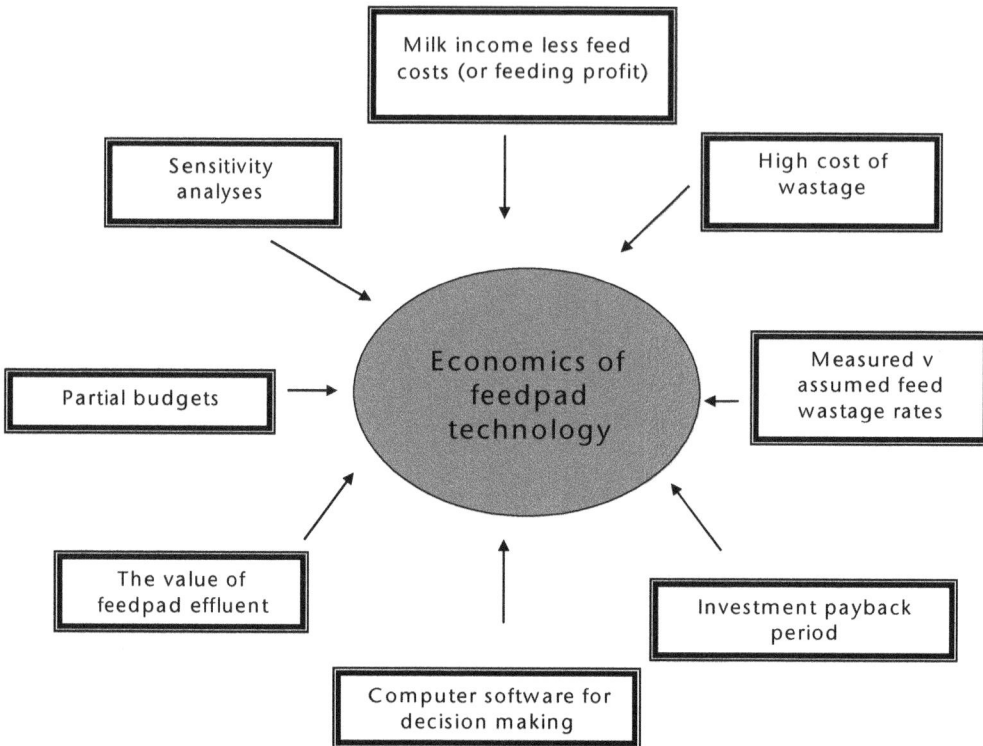

Figure 11.1 Some of the key considerations for assessing the economics of feedpad technology.

Dairy farming is a business so the bottom line from any management decision is whether it was profitable. Feedpad systems are not cheap, compared to other investments in farm feeding practices and can cost $60 000 to $100 000 for a farm with 250 milking cows. Therefore it is essential that both the estimated costs and additional returns are quantified before making any purchases of materials, contractors or specific machinery. This chapter provides direction in understanding some of these key financial outlays and the extra income they can produce, to assist with the ultimate decision 'Is a feedpad system a good investment for my dairy enterprise now, in the future or not at all?' Topics addressed in this chapter are a simple and useful measure of day-to-day farm performance is milk income less purchased feed costs; minimising the high costs of wastage of supplementary feeds, a major incentive for investing in feedpad technology; computer software is available to assist with decisions in feedpad investment; feedpad effluent has an added benefit of recycling farm nutrients which can reduce inputs of inorganic fertilisers for grazed pastures and other forage crops; and the best way to assess the potential value of feedpad technology is through partial budgeting (see Figure 11.1).

Milk income less feed costs

The profitability of dairy farming is largely driven by the cost of feed, the returns for milk and the relativity of the two. Feed costs generally comprise 50–60% of total production costs. Any detailed assessment of farm profits should also consider other variable costs, such as herd and shed costs, overhead or fixed costs, such labour, depreciation and finance costs, and other farm returns, such as stock sales and changes in stock and farm value.

As feed costs and milk returns are the major profit drivers, these deserve detailed attention in any discussions on feedpad economics. The calculation of milk income less feed costs provides a simple measure of 'feeding profit'.

Feed costs

Feed costs include all the costs to grow and purchase feed for the dairy herd. In northern Victoria, 20 or 30 years ago when grazed pastures comprised virtually all of the feed supplies and irrigation water was readily available, feed costs typically made up only 30–40% of production costs. The recent drought years (2002/3 and 2006/7) have increased this proportion to 60% or more. No longer is homegrown feed always the cheapest feed, as the key inputs (irrigation water, fertilisers, seed) have all risen, making homegrown feed a lot more expensive. Therefore supplements have now become important to provide potentially more cost-effective feeds. Purchased supplements require efficient feed handling and feeding out systems to reduce their unit costs.

In addition, the sources of grazed forages have changed over recent years as farmers seek to produce increased and higher quality forages with limited irrigation water. The choices have been perennial and annual pastures and perennial and annual forage crops. No farming system is without risk. Risk profiles are obviously different, but equally as important, in rain-fed dairy systems.

Different farming systems have different types of exposure to feed cost risk (Gibb 2009). For example, the traditional perennial pasture system is highly reliant on availability and unit cost of irrigation water. In low water allocation years, farmers are very exposed to trading in temporary water markets, unless there are other water sources, such as groundwater; dairy systems that depend on high levels of inputs (grains, hay) are exposed to fluctuating market prices; and systems that rely on high levels of machinery for harvesting and/or feeding out conserved or purchased fodder introduce a new fixed cost risk, with capital invested in machinery and infrastructure and associated depreciation. Farmers often manage this risk by increasing scale to spread the fixed costs over more output, but they must be aware of their own particular risk profile.

Changing farming systems involves changing risk profiles. For example, when growing more fodder to reduce the risk of purchased fodder price, farmers are more exposed to crop failure, so contingency plans to overcome such shortfalls in feed supplies; climate, even when irrigation provides flexibility with water supplies, for example cold weather or prolonged rainfall events; quality and palatability of homegrown feeds; timeliness of supply of these feed inputs to match herd requirements; wastage when conserving, storing and feeding out; time input to prepare the fodder for feeding out; infrastructure to use fodder crops; and weather during storage and feed out.

Milk returns

Milk price is the biggest risk of any dairy farming system (Moran 2009). On most farms it contributes over 90% of farm returns so small changes in milk returns can have dramatic effects on farm profitability. The annual average milk return is the most important in determining profitability, hence farming system. Within any one year, there are variations between seasons, monthly averages as well as variations in spot price. In Chapter 2, Figure 2.4 plots the annual milk returns ($/ kg milk protein and fat) for dairy farmers supplying Murray Goulburn Dairy Cooperative over the 20 years from 1990 to 2009. From this dataset, Gibb (2009) calculated a 15% fluctuation around the trend line. That was until 2008 when milk price jumped to 55% above the trend line, after which the 2008 price returned to the trend line. After adjusting for annual inflation, Gibb (2009) concluded that this trend line actually changed from positive to negative, meaning that 'real' milk prices have been steadily decreasing over the last 20 years. Furthermore, milk processors predominantly supplying export markets (such as Murray Goulburn) had more variable prices than those supplying domestic markets, since they were exposed to the uncertainties of global dairy prices and currency exchange rates.

The above discussion is necessary for farmers to assess the risk of their particular production system. Without some prediction of their milk returns, they cannot plan their long-term investment strategies objectively nor their day-to-day farming programs. Farmers can make long-term decisions such as changing their calving patterns or developing farm infrastructure such as feedpads; medium-term decisions such as varying their pasture crop species mix, or their pre-purchases or contracts for cereal grains, by-products, hay or silage. Selecting the most appropriate milk processor to match their supply pattern and cost structure; and short-term decisions such as varying their irrigated water application or their level of grain feeding on a day-to-day basis.

The high export milk prices in 2008 (from Figure 2.4) led to many dairy farmers making the short-term decisions to intensify their production systems. In other words, they invested in infrastructures to move along the spectrum of feeding systems from low to high feed inputs (see Table 2.4 in Chapter 2), as described by Little (2009a). This in turn, led to a rapid increase in the number of enquiries (see Figure 1.2 in Chapter 1) and the actual construction of feedpads, particularly in northern Victoria. The unpredicted sharp decline in 2009 export milk prices created considerable uncertainty about the wisdom of such decisions.

For every dairy farmer, there is no right or wrong system, only the most suitable for their farm. There are highly skilled farmers operating throughout the full range of systems; business returns for the top 10% of each system are not all that different. Having the diversity of operators and systems is one of the best defences against the vagrancies of agriculture.

Calculating feeding profit

Profitability can be expressed in many ways. In this context, it is simply milk income less feed costs. Feed costs are the sum of homegrown and purchased feeds. Calculating the costs of homegrown feed, however, requires detailed analyses of all inputs but of more importance, an estimate of the amount of feed grown. This is not easy because a farmer would rarely have any idea of how much feed his farm produces. A simpler approach is to just consider the costs of purchased feeds. Although this is less accurate, it can still provide a useful guide to the cost efficiency of any dairy farming system.

Ideally feed costs should be the cost of the feed 'down the cow's throat'; that is, the cost of purchasing or sourcing the feed, then storing, processing, offering the feed to the stock and finally accounting for the feed sourced and eventually offered but not consumed (these are the losses during storage and the wastage following feed out). Such a series of calculations are rarely, if ever, undertaken. Usually 'feed costs' are limited to the purchase cost of concentrates or forages or an estimate of the costs of inputs and harvesting homegrown feed.

Examples of calculating feeding profit

The following example uses data from previous chapters which have been incorporated into the computer model MIFC10 (which is available free from the author: john.moran@dpi.vic.gov.au or jbm95@hotmail.com). This example calculates the milk income less purchased feed costs for three herds on milking cows, producing on average 20, 25 or 30 kg/cow/d of milk. The ME requirements of these herds are broken down in Table 6.1. In this scenario, each herd consumes 6 kg DM/cow/d as grazed perennial ryegrass pastures (containing 11 MJ/kg DM of ME, 16% CP and 40% NDF) and is fed the remainder of their diet in a PMR. The feeds available, their nutritive values (from Table 7.2) and purchased costs are presented in Table 11.1. Milk returns were 34 c/kg. Total nutrient intakes, purchased feed costs, milk returns and the resulting profits (expressed as milk income less purchased feed costs) are presented in Table 11.2.

Clearly, despite higher feed intakes hence higher purchased feed costs, the feeding profits increased as cows produced more milk. This does not necessarily mean that higher yielding cows always generate more total profit. The changing

Table 11.1 Nutritive contents and purchase costs of feeds available for PMR.

Feed	Price ($/t fresh)	DM (%)	ME (MJ/kg DM)	CP (%)	NDF (%)	c/kg DM	c/MJ ME	c/kg CP
Barley	250	88	12	11	20	28.4	2.4	258.3
Canola meal	350	90	12	37	30	38.9	3.2	105.1
Oaten silage	70	41	9	10	60	17.1	1.9	170.7
Lucerne hay	230	88	9	19	45	26.1	2.9	137.6
Brewers grain	30	28	11	22	55	10.7	1.0	48.7

profit margin depends on the marginal cost and marginal return of each additional unit of feed input. In the above example, each additional MJ of ME cost 2.7 cents and produced an additional 6.5 cents in extra milk, so naturally the more milk the higher the feeding profit margins. However this is purely based on the cost of additional feed. As already mentioned, this must pay for all the other operating

Table 11.2 Formulation of PMR, total feed costs and feeding profits arising from three herds supplemented on a feedpad.

	Herd 1	Herd 2	Herd 3
Milk yield (kg/cow/d)	30	25	20
ME requirements (MJ/cow/d)	203	177	150
Grazed pasture intake (kg DM/cow/d)	6.0	6.0	6.0
Supplement intake (kg fresh/cow/d) of:			
Barley	7	6	4.5
Canola meal	2	1	0.5
Oaten silage	2	1	1
Lucerne hay	3	3	2
Brewers grain	3	3	3
Total DM intake	**18.3**	**16.1**	**13.7**
ME intake (MJ/cow/d)	202	177	151
CP content (%)	17	16	16
NDF content (%)	35	35	36
PMR cost ($/t DM)	274	267	253
Milk returns ($/cow/d)	10.20	8.50	6.80
Purchased feed costs ($/cow/d)	3.37	2.70	1.95
Milk income less purchased feed ($/cow/d)	**6.83**	**5.80**	**4.85**

costs (such as herd, shed, other feed inputs such as pasture costs) and the overhead costs (labour, office administration, finance, capital and/or depreciation). Granted many of these would not change purely because cows produced more milk, as they would if, say, herd size increased to increase farm milk output. As cows are fed better, however, the unit feed costs often increase.

The above example falls into the trap of assuming linearity of input/output responses, as discussed in Chapter 6 (see section 'The lactation cycle'), in which Figure 6.6 highlights the concept of decreasing marginal milk responses. Until sufficient research has been undertaken, data are just not available to develop such a figure for milk responses to PMRs. It is highly likely, however, that increases in intake of the same PMR will not result in corresponding linear increases in milk production. Therefore the theoretical data presented in Table 11.2 must be tested in the 'real world' to validate the results. How can this be done without building the facilities and investing in the infrastructure of feedpad technology? That is the role of the research scientist and publicly funded research institutes. In the meantime, we can only use the computer models developed by such dairy industry specialists with all their inherent inaccuracies. This includes knowing that they are likely to overpredict, to some degree, theoretical milk responses hence any predicted profit margins to intensification of dairy production systems.

The high costs of feed wastage

There is little documented information available to quantify feed out losses (wastage) under different feeding practices. Wastage occurs through aerobic spoilage in silage, which includes losses in DM and nutrients occurring when the silage is exposed to air while in the storage, during its removal from the stack and during feed out; losses during storage can amount to 15–20% inside horizontal silage bunkers or 5–10% inside individually wrapped round bales (Moran 1996); losses during feed out can amount to 2–5% (Moran 1996); stock trampling and lying on the feed; stock urinating and defecating on the feed; feed that the stock refuse to eat due to it becoming mixed with mud (if wet), moulds, or other unpalatable contaminants; the filtering of finer particles of feed mix (such as finely ground cereal grain or small leaves of conserved feed) onto the feeding surface where they are more likely to be rejected; and errors in assumptions of average bale weights and quantities of grain being bail fed

The degree of feed wastage depends on the type of feed (hay versus rolled grain); the type of hay (clover versus grass) and the degree of leafiness of hays (due to leaf shatter); method of feeding out (for example, grain fed out in troughs in paddock v in the milking shed); feed out equipment (for example, hay fed out on ground versus in hay rings); maintenance of equipment (such as routinely testing grain feeders); the condition/palatability of the feed (such as fresh versus mouldy silage); weather (when feeding out in paddock/laneways versus under cover); feed out surfaces (such as pasture versus dirt versus concrete surfaces); and operator skills in presenting the feed in the paddock of on the feedpad.

Estimates of losses from inefficient feed out systems

From a study with beef cattle fed round bales of hay, wastage rates increased as more was fed out in the paddock, from 11% (when cows were fed 9 kg/cow/d) to 34% (when offered 72 kg/cow) compared to only 5% when the hay was fed out in hay racks. In another study when chopped silage was fed to dairy cows, losses were 23% when fed out in the paddock compared to 6% when fed in troughs.

Stevens and Flatfoot (2005) monitored losses of precision-chopped whole crop cereal and pasture silages when fed out to dairy cows in the paddock in spring and autumn, recording DM losses varying from 9% in pasture silage fed along a fence line to 19% in cereal silage fed in the middle of the paddock. Seasonal influences were not apparent. They recorded wastage rates of up to 38% in silages containing stones (8% of silage DM). As expected, they reported utilisation of the silage to be related positively to the amount fed out, as quantified by the density of the silage on the ground.

Figure 11.2 Hay rings can still lead to large wastages during feed out.

Stockdale (2010) recently reviewed the published scientific literature on wastage of hay and silage when fed out to sheep and beef cattle directly onto the pasture, in hay rings or chopped and in troughs. Wastage rates ranged from 1% to 77% of that fed out, but from 15 datasets per mean value, losses average 31% (ranging from 5% to 77%) when fed out directly onto pasture compared to 5% (ranging from 1% to 9%) when fed via hay rings or into troughs.

In the light of limited data from actual studies involving dairy cows, advisers can only make best 'guesstimates' of anticipated losses for different feeding out scenarios. Until such data become available, the following have to be accepted as industry 'norms' on which to base assumptions to calculate the costs of inefficient feed out systems (Dairy Australia 2007b; McDonald *et al.* 2008).

- Feeding on ground   30% or more
- Hay ring   10–20%
- Conveyer belt   up to 15%
- Concrete troughs   5–8%
- Concrete pad and free stall shed   less than 5%

Calculations of financial losses and investment payback periods

During 2006 and 2007, purchased lucerne hay increased in cost from $200 to $500/t (see Figure 2.6 in Chapter 2). Pasture and cereal hay was valued at $300/t in 2008 but this decreased to a predicted $200/t in spring 2009. The cost of purchased silage would also reflect such wide variations in hay prices. This will have a dramatic effect on estimates of financial losses during feed out for a large dairy herd, thus will impact on farmers' decisions to invest in feedpad technology.

For a herd with limited grazed pastures, milking cows can be fed up to 3 t/cow of hay, or 6 t/cow of fresh silage over a full lactation. Table 11.3, using 'typical' feed out losses, presents a range of financial costs of inefficient feed out systems. Three values were assumed, for 30% for high (paddock feeding), 15% for moderate (using hay rings) and 5% for low (on feedpads) wastage rates.

Using hay rings, rather than feeding hay out directly in the paddock, could save $90–$120/cow/yr depending on hay prices. Further investing in a concrete feedpad could increase these savings to $150–$200/cow/yr.

Table 11.3 Calculated losses in total hay fed out for various levels of wastage for cows each fed 3 t hay over a full lactation.

Wastage	Feeding out	% loss	Loss/cow (kg/cow)	Cost at $300/t ($/cow)	Cost at $200/t ($/cow)
High	Paddock	30	900	252	180
Moderate	Hay rings	15	450	135	90
Minimal	Feedpad	5	150	45	30

Dairy Australia (2007b) has classified feedpads into four categories (see full details on Table 3.1 in Chapter 3). Briefly, with estimated cost/cow and feed wastage rates, these are:

1. Temporary and relocatable: <$50/cow and >30% wastage.
2. Semi-permanent: $50–100/cow and 15–30% wastage.
3. Permanent and basic: $100–200/cow and 8–15% wastage.
4. Permanent and maximum control: >$200/cow and 5–8% wastage.

Theoretical scenarios for a typical dairy farm in southern Australia with limited pasture for grazing would be a herd of 250 milking cows supplemented with a mixture of hay, silage, cereal grain and some by-products, totalling 3 t DM/cow/yr. The average cost of the PMR, similar to those fed to cows in Table 11.2 would be $270/t DM, making a total annual supplementary feeding bill of $810/cow or $202 500 for the farm. To improve the feeding system, it is assumed that there is no salvage value of previous feedpad facilities; in other words, each new improved facility was constructed on a greenfield site on the farm.

Table 11.4 presents the number of years required to pay back the investment cost of an improved feedpad based on annual savings from reduced feed wastage. This table does not account for the additional cost of interest on expenditure during the period of time to repay the investment. In most cases, the investment cost was returned within two years, except when upgrading a permanent basic feedpad to a permanent one with maximum control, such as a free stall shed. As the number of years payback period is directly related to the annual feed loss and if the farmer was only feeding 2 t DM/cow/yr of supplement, this period would increase by 33%.

Feed out losses are invariably quantified in terms of percentage of the fresh feed (or sometimes the DM) that is offered to the stock, which they do not consume. Cow performance depends on the quantity of feed nutrients consumed,

Table 11.4 Payback period for investing in improved feedpad technology for a herd of 250 cows on four feedpads of different level of sophistication.

Feedpad category	Assumed total construction cost ($000)	Assumed feed wastage rate (%)	Annual feed loss ($000)	Payback period from Feedpad 1 (years)	Payback period from Feedpad 2 (years	Payback period from Feedpad 3 (years)
1	10.0	30	60.7	–	–	–
2	18.7	22	44.5	1.2	–	–
3	37.5	11	22.2	1.0	1.7	–
4	62.5	6	12.2	1.3	1.9	6.2

usually dietary energy or protein. Ideally the costs of feed wastage should take into account the composition of the feeds wasted hence not consumed. This requires some estimate of the composition of the feeds left behind in the paddock or on the feedpad.

This will depend on the type of feed and how the wastage occurred. For example, if a mixed ration was fed out and the fine particles of concentrates filtered out onto the ground, any economic analyses including losses arising from feed wastage should place a higher value per tonne of feed lost than on the feed actually fed out. The same argument would apply to most hay wastages on dry surfaces, because leaf shatter usually occurs hence the hay lost has a higher nutritive value than the hay fed out. It is virtually impossible to place an accurate monetary value on feed wastage, but such costs (as in Table 11.3 and 11.4) are invariably underestimated.

Clearly considerable money is likely to be wasted on dairy farms that depend on inefficient feed out facilities, particularly as unit feed costs rise and grazed pasture becomes more limited as a result of unreliable rainfall or irrigation water supplies and increased stocking rates. These scenarios do not even consider the opportunities to feed young stock on the improved feedpad. Herd size is also an issue because the investment costs in Table 11.4 are only for constructing the feed out area and do not take machinery for handling the PMR or for effluent management. The larger the herd size up to a point, the lower these investment costs per cow, provided additional feedpad infrastructure can be optimised.

Computer software to assist with making feedpad investment decisions

A new computer program, called Feedout Checkout, has recently been developed specifically to analyse whether the financial benefits of reduced feed wastage and improved cow performance are sufficient to justify the extra capital investment on a feedpad and associated machinery (Leddin and Armstrong 2009). The program predicts investment payback periods using a partial budget approach (see 'Using partial budgets to plan farm development' below).

Data are entered into three sections as follows:

- Capital costs
  - Type of feedpad (using Dairy Australia 2007b categories)
  - Feedpad cost, just for the feed out area ($/cow and total $)
  - Effluent upgrade (total $)
  - Upgrading silage cart or mixer wagon ($ with estimate of how many years it will last, namely five or 10 years)

- Upgrading tractor and front-end loader ($)
- Other capital costs, such as cow loafing or feed storage areas or a cow cooling system ($). This cell can also be used to include salvage value of old equipment that can be sold, which are entered as negative values.
- Total capital cost (all of above)
- Cows and feeding
  - Size of milking herd (cows)
  - Total amount of feed offered on feedpad (t DM/cow/y). This can be generated from a separate feed ready reckoner where details of all feeds and their costs are inserted.
  - Average cost of feed generated from the feed ready reckoner ($/t DM). As this computer program evaluates the feedpad investment over 10 years, it is best to run the program with some different feed costs to take into account estimates of feed price increases.
  - Reduced wastage, using the range of estimates listed above (%)
  - Improvement of metabolisable energy (ME) supply to cow due to improved rumen metabolism, with a maximum of, say, 5% if changing to a mixer wagon incorporating both forage and concentrates in a PMR (%).
- Other costs and benefits
  - Increase in fuel costs, due to increased tractor usage ($/t DM and $/yr)
  - Increase in repairs and maintenance due to increased usage of machinery ($/t DM and $/yr)
  - Change in labour requirement (hr/d) and labour cost ($/hr)
  - Other operating costs, such as managing mastitis or other animal health issues ($/yr)
  - Other benefits, see below for such benefits ($/yr)
  - Interest rate, for any borrowed finance (%)

The partial budget approach does not consider a number of important benefits arising from installing a feedpad although these can be included under 'Other benefits' above. These include such benefits as improving the capital value of the farm; increasing farm-stocking capacity; improving per cow production; reducing pugging of pastures during wet periods; and more efficient heat-stress management.

It is not always easy to place a monetary value for such benefits.

The program then calculates the following measures to assist with the investment decision:

- The years to pay back the investment. This is the time taken for the return to pay for the investment's purchase. This occurs when the cumulative net cash flow becomes positive. It is not a measure of economic or financial benefit, simply the time taken to remove the debt and regain positive cash flow. Time estimates are provided either before interest or after interest, to cater for situations where the investment is funded from cash reserves or where all the money is borrowed.

- Net present value (NPV). This is the additional wealth compared to an alternative investment. The program uses a discount rate of 10%, meaning that the alternative investment could earn a 10% return. If the NPV is positive, the feedpad is a better investment than alternative options that can earn 10% return. If the NPV is negative, a feedpad may not be the best investment in the current circumstances.

- Internal rate of return (IRR). This is the average annual return on the capital invested over the estimated life of the facility (that is, 10 years) and gives a clear guide as to whether the project is worth undertaking. The higher the IRR, the better the value and return on investment. A more risky or uncertain investment requires a higher IRR to justify the project. The IRR is also a measure of the extra return to the additional capital invested in developing the feedpad. If it is greater than could be earned in other farm development projects, then it is a good investment. It is actually the discount rate that will result in zero NPV to the investment, that is the maximum interest rate that farmers could afford to pay for the feedpad development if they borrowed all the funds and were still able to refund the capital borrowed, plus interest. If it is less than the interest rate for borrowing money, then it is obviously not a good investment.

Leddin and Armstrong's (2009) program provides a more robust assessment of investment in feedpad technology than the somewhat simplistic approach presented in Table 11.4, but its accuracy does depend on considerably more assumptions. These include the improvements in the utilisation of dietary ME, any increases in fuel usage, machinery repairs and maintenance and labour requirements, as well as estimates of changes in other operating costs and whole-farm benefits. Since many of these may be unknown, these variables can be placed at zero during the first run of the computer model, after which the model can be run again, this time with specific estimates, to assess what effects they are likely to have on the bottom line, namely the investment payback period, NPV and IRR.

The value of feedpad effluent

Spending money on an efficient effluent system is more likely to be a good investment if it can maximise the benefits of recycling nutrients. Between 80–90% of the inorganic nutrients ingested by cows in their feed are excreted in their dung and urine. Since these nutrients originate from herbages stock consume, it is not surprising that manure contains all the nutrients required by plants.

Each 500 kg dairy cow produces on average, 40 kg/d of raw manure or 4.2 kg/d of solids (SIRIC 2002), which supplies the following plant nutrients:

- Nitrogen (N): 0.225 kg/d

Table 11.5 Financial benefits from recycling effluent nutrients for a 250-cow herd spending half the time on a feedpad.

Nutrient	Daily excretion (g/cow/d)	Nutrients collected (kg/herd/d)	Equivalent in fertiliser (kg/d)	Fertiliser cost ($/t)	Daily fertiliser value ($)
Nitrogen	225	28.1	61.1	Urea	32
Phosphorus	47	5.9	66.8	Single super	18
Potassium	145	18.1	36.2	Potash	31

- Phosphorus (P): 0.047 kg/d
- Potassium (K): 0.145 kg/d

The nutrient contents of fertilisers (and their costs in AUS$ Dec 2009) partially replaced by effluent recycling are:

Nitrogen   Urea (46% N) @ $517/t

Phosphorus Single superphosphate (8.8% P) @ $275/t

                  Di Ammonium Phosphate (DAP) (18% N, 20% P) @ $540/t

Potassium   Muriate of potash (50% K) @ $850/t

Table 11.5 presents data for a 250-cow herd with 50% of their effluent being collected on a feedpad and recycled back onto the farm pastures. The calculations assume that the value of the nutrients in the effluent is only obtained if they are recycled back onto the paddocks.

The full value of the recycled effluent is then $81/d or say $16 200/yr if the herd spends 200 d/yr utilising the feedpad.

On a DM basis, dairy cow manure contains 2–4% nitrogen (N), 0.3–0.8% phosphorus (P), 1–3% potassium (K), 1–2% calcium (Ca) and 0.4–0.9% magnesium (Mg). Most of the P is in the solid dung while most of the N is in the urine.

Although manures contain relatively low percentages of plant nutrients, they can be safely applied in large quantities to the soils. For example, with effluent containing 75% moisture, applying it at a rate of 25 t/ha would provide about 175 kg N/ha, 50 kg P/ha and 100 kg K/ha. The availability of these nutrients for plant growth varies and may extend over a number of years. Nitrogen becomes available over a 3–4 year period with 40–60% being available in the first year. Phosphorus follows a similar pattern while potassium and sodium chloride are all readily available within the first year.

Ideally, manure should be incorporated into the soil immediately following application. As well as minimising odour and fly problems, this also reduces loss on N as ammonia drastically. Pasture is a versatile crop in that manure can be spread at almost any time of the year. Spreading effluent on pasture surfaces can, however, cause problems by reducing palatability of grass in the short term.

Using manure for crop and pasture production is no more 'natural' than using manufactured fertilisers. It still represents a net transfer of nutrients from one place to another within the farm, instead of from the fertiliser factory to the farm. One major benefit over inorganic fertilisers is the improvement of soil organic matter levels through using manures which would increase microbial activity in the soil and nutrient recycling.

Using partial budgets to plan farm development

Partial budgets provide planning and decision making frameworks to compare the costs and benefits of alternative farm practices, by focusing on the changes in income and expenses resulting from that alternative. Thus all aspects of farm profits that are unchanged by the decision can be safely ignored. In a nutshell, partial budgeting allows farmers to get a better handle on how a decision will affect the overall profitability of their farm enterprise (Moran 2009).

Planning includes taking an inventory of farm resources, devising alternative uses of these resources and choosing the best alternative. By employing budgeting principles, the farmer can compare costs and returns of a range of alternative actions. Ideally, farmers should aim to choose a course of action that most clearly matches their long-term goals.

The key principals of budgeting are to prepare for the unexpected and to measure past performance against future profits. Good budgets begin with specific measurable goals that provide clear direction to all involved in the business.

The success of the partial budget is reliant on its predictive accuracy, which depends on the accuracy of the information and estimates it contains. Farmers need to collect factual data about the proposed change and provide reasonable estimates of such items as future prices, yields and gains. Factual information includes current costs of the production inputs, costs of capital, current commodity prices or other items pertinent to the change. The process may or may not account for time lags between when an investment was made and when it realised the full return.

It is difficult to generate estimates of the unknown, particularly prices. Farmers must then estimate yields and prices to get some idea of what these returns will or will not be.

Steps for a partial budget

There are eight steps to a successful use of partial budgets analyses as a decision-making tool. Each step serves a specific, unique purpose and is vital to an accurate, meaningful analysis. These steps are:

1. State the proposed change. It is important to have a clear understanding of exactly what alternative is being analysed.

2. List the added returns. Identify new revenue streams or increasing existing streams.
3. List the reduced costs. Identify the general areas where a choice might lower expenses.
4. List the added costs. In the situation of capital purchases, a depreciated cost must be claimed annually, not the total purchase cost.
5. List the reduced returns. Will revenues be decreased as a result of choosing a particular alternative? Will it decrease yields?
6. Summarise the net effects. Once the individual positive (steps 2 and 3) and negative (steps 4 and 5) aspects of the alternative have been identified, they should be aggregated to determine a total cost and total benefit of the alternative. The net benefit is found by subtracting total costs from total benefits. If the result is positive, then that alternative may have some economic advantages. If it is negative, however, the business would be better off staying with the current situation or analysing a different alternative.
7. If capital is invested in the development, it is useful to calculate the return on that marginal capital. The question must then be asked, 'What is an acceptable return on this marginal capital?' It certainly should be more than current interest rates for investing that money in other ventures, either off-farm or on-farm. In making a value judgement, advisers often seek a value of at least 25–30% to take into account the risk and uncertainty of the calculated net benefits (step 6) actually being realised.
8. Consider non-economic and other factors. These must be taken into account but are difficult to quantify. If the alternative involves increasing herd size, it is important to consider the existing farm infrastructure, such as the capacity of the dairy and the milking sheds; if these are already stocked to near full capacity, they may have to be enlarged, thus requiring additional capital investment. Other considerations may include the social aspects of having less labour on the farm, increased or decreased leisure time, the need for increased or specialised knowledge and safety and/or ease of use of new equipment. Placing a monetary value on some of these subjective factors may be nigh impossible.

Example of a partial budget

Busby *et al.* (2007) presented an example partial budget to decide whether to purchase a mixer wagon. The proposed change was to intensify the current production system to make best use of the new purchase.

The associated marginal capital could then include:

- mixer wagon with scales
- feed mill
- front-end loader

Table 11.6 Components of a partial budget to assess the benefits of purchasing a mixer wagon.

Positive effects	Negative effects
Added returns • Increase in cow and herd milk production • Increase in milk fat and protein contents • Sale of old equipment	Added costs • Extra feed • Extra labour • Extra fuel for tractor • Extra repairs and maintenance on tractor • Finance costs (interest or lease plus residual payments) • Nutrition consultant fees • Feed analyses • Computer programs
Reduced costs • Reduced feed wastage	Reduced returns Nil
Net benefit	
Return on marginal capital	

- commodity shed or bays for storing bulk feed purchases
- additional fencing, feed and water troughs
- effluent management.

The components of the partial budget are presented in Table 11.6.
The net benefit, that is the profit or loss, would be calculated from:

$$\text{(Added returns + reduced costs)} - \text{(added costs + reduced returns)}$$

The return on marginal capital is calculated from:

$$\frac{\text{Net benefit}}{\text{Marginal capital}}$$

Sensitivity analyses

When varying the level or costs of farm inputs and monitoring their effects on outputs, it is apparent that some inputs have a large influence on profits while others have very little influence. In addition, farm inputs often do not operate in isolation, but interact with others. This can be quantified using a sensitivity analysis.

Greatest attention should be given to the high impact variables to ensure they are predicted most accurately. For example, unit milk returns and unit cereal grain prices would be the two most important variables to include in a sensitivity analysis on profits from changing feeding systems. These analyses involve repeating the partial budget with differing values for milk returns and grain prices, say by varying them 10% above and 10% below their assumed value in the original budget.

12

Guidelines and legislation

This chapter describes the key legislation required for dairy farmers to consider when establishing feedpads in Victoria.

The main points in this chapter:

- Rules on the development and expansion of feedpads are often ambiguous and so can lead to disputes. One major question is: when does a broadacre farm becomes an intensive animal enterprise or a feedlot?
- Regulatory requirements can vary between states.
- Planning conditions usually govern buffer distances to neighbouring houses, waterways, groundwater bores, roads and towns, while structures need to conform to building standards.
- Planning permits and Environmental Protection Authority (EPA) works approval may be required depending on the size of the proposed feedpad.
- Within Victoria, there is a range of statutory bodies in addition to the EPA that might need to become involved. A list of relevant State Government Acts of Parliament are included in this chapter
- A series of concerns expressed by local government, dairy company and Department of Primary Industries is included in this chapter

Rules on the development and expansion of feedpads are often ambiguous and can lead to disputes. One major question is: when does a broadacre farm become an intensive animal enterprise or a feedlot? Under drought conditions or where inclement weather is common, farmers are encouraged to develop structures and systems for supplementary feeding, but local planning ordinances must be followed.

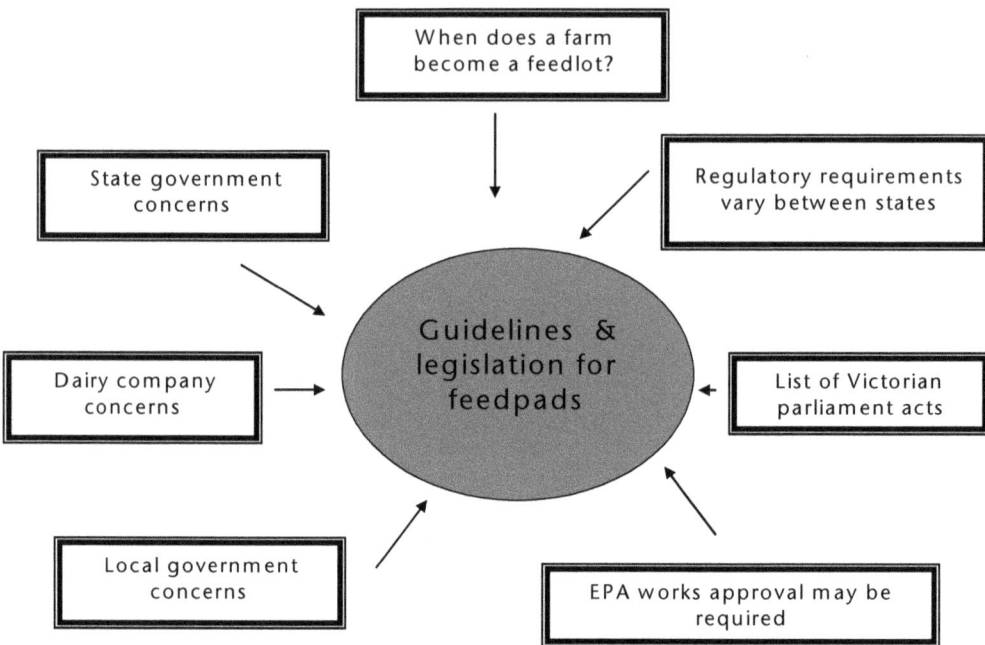

Figure 12.1 Some of the key concerns about guidelines and legislation when developing feedpads.

Regulatory requirements for feedpads can vary between states. Planning conditions usually govern buffer distances to neighbouring houses, waterways, groundwater bores, roads and towns, while structures need to conform to building standards. Waste management conditions are commonly dictated by the scale of works which usually set requirements for the storage and land application of effluent and manure.

Most of the concerns that planning authorities have with intense dairy operations relate to odour, noise, light access, heavy vehicle movement, loss of property value and amenity, the fate of wastes and the use of medications. Some animal welfare groups campaign against intensive animal housing and occasionally national animal welfare agencies become involved in the regulation of such facilities.

Guidelines for feedpad development

A set of comprehensive guidelines was developed for northern Victoria by the Shepparton Irrigation Regional Implementation Committee (SIRIC 2002). These were updated in 2010 (O'Keefe *et al.* 2010). The aim of these guidelines is to provide a rational process to assist consultants, farmers and regulatory authorities in the planning, development, construction and management of dairy feedpad systems to ensure balance between dairy production, livestock health, environmental protection, sustainability and minimal impacts of neighbours by:

- Providing background information for those considering a feedpad such as the benefits, required management changes, options, etc.
- Providing practical information on a range of aspects relating to various feedpads.
- Providing a rational process for those who have decided to proceed with a feedpad to help ensure a quicker and less difficult process.
- Promoting the establishment and management of feedpad systems that are best managed practice and are environmentally responsible.
- Helping prevent potential adverse impacts on waterways by affirming that any waste generated should stay on farm.
- Assisting those implementing or managing a feedpad to meet their legal and social responsibilities thereby upholding the high regard the community currently has of the dairy industry.
- Providing avenues to access further information and assistance.

Legislation and planning requirements

When considering a new feedpad or expanding an existing one, it is important to bear in mind the legislation that might apply, such as whether a feedpad is an allowable use having regard to the zoning of the site and other relevant planning provisions; whether a planning permit will be required; whether an Environmental Protection Authority (EPA) Works Approval will be required; and which other government authorities may be involved in the assessment of any planning permit application and the information they might require.

A planning permit is required if the proposed feedpad is considered to be a cattle feedlot with a capacity exceeding 50 head; the site is located outside specific locations, such as water supply catchments; an overlay applies to the site. These generally relate to issues affecting the environment such as whether the land is subjected to inundation, heritage and land and site issues. You must also apply for a permit if native vegetation will be removed; a building is to be constructed; and where the cattle are supplied with the majority of their feed requirements while on the feedpad. In this case, the feedpad will alter the type of stock management from extensive to intensive animal husbandry (see glossary for complete definition of intensive animal husbandry). An additional planning permit is required if the proposed feedpad is considered to be a cattle feedlot with a planned capacity exceeding 1000 head. An EPA Works Approval is required if the proposed feedpad is considered to be a cattle feedlot with a capacity exceeding 5000 head. Even if the proposed feedpad will hold less than 50 head, the

Figure 12.2 It is important to check all relevant legislation when constructing new feedpads.

recommendations published by SIRIC (2002) and O'Keefe *et al.* (2010) should be followed.

Within Victoria, there are a range of statutory bodies in addition to the EPA that might need to become involved. Just to provide some indication of the government legislation that may influence feedpad development, the following Acts (and their year of legislation) can be listed:

- *Environmental Protection Act* (1970)
- *Aboriginal Heritage Act* (2006)
- *Occupational Health and Safety Act* (2004)
- *Environment Protection and Biodiversity Conservation Act* (1999)
- *Planning and Environmental Act* (1997)
- *Catchment and Land Protection Act* (1994)
- *Water Acts* (1989 and 2002)
- *Flora and Fauna Guarantee Act* (1988)
- *Code of Practice of Cattle Feedlots*
- *Interim Guidelines for Control of Noise from Industry in Country Victoria.*

Details of these Acts are outside the scope of this book and more details have been provided by O'Keefe *et al.* (2010), as well as the rather complex application process required within Victoria.

The *Environmental Protection Act* (1970) describes the relevant obligations for a proposed feedpad owner who must contain and reuse all the manure deposited on dairy tracts, underpasses and the feedpad complex (most commonly applied back to the pastures and crops); ensure that manure does not enter surface waters (including dams, impoundments, rivers, creeks and all waterways where rainfall is likely to collect); ensure that manure does not enter groundwaters either directly or though infiltration, such as excessive seepage from ponds or from stockpiled manure; ensure that manure does not contaminate land (for example, regular applications of manure on a small land area may result in excessive nutrient levels in the soil); ensure that offensive odours do not impact beyond property boundaries; and apply for an EPA Works Approval if the proposed development is considered to be a cattle feedlot with a capacity exceeding 5000 head, after which the Feedlot Code comes into play.

As municipal (or local) government agencies are generally responsible for administrating or enforcing planning schemes, they assess the applications for feedpad planning permits. They are also responsible for the monitoring and enforcing compliance with planning permit conditions and schemes. Other Victorian Government agencies that may have a statutory or voluntary obligation to become involved include the Environmental Protection Authority, the Department of Primary Industries, any Catchment Management Authority, Rural and Water Authorities, the Department of Sustainability and Environment, and Vic Roads.

Concerns of government agencies in Victoria

Dairy Australia (2008b) summarised the range of concerns of various Government agencies in Victoria in feedpad development comprehensively. These include:

Local Government Planning Approval – buffer distances to neighbouring properties and residences; buffer distances to watercourses and to roads; noise; odour; lighting; road access and slight distances; stormwater control and effluent management; building standards; hours of operation; aesthetics; and vegetation management/clearance of trees.

Dairy Company Quality Assurance – milk harvesting regulations; distances to the milk room from the facility; distances to effluent-management system; and quality control for milk harvesting.

Department of Primary Industry concerns, including animal welfare; effluent management; use of manure and soiled bedding; environmental impact; biodiversity; access to and from feedpad for animals; vehicle access to feedpad for delivery of feed and removal of manure; aspect; shelter and shade; water supply; feed storage and baiting for vermin; cleaning feedbunk or feed out area; control of runoff to and from pad; control of effluent – storage and re-use; harvesting and storage of manure; visibility and aesthetics; stages of construction.

13

The future role for feedpads in Australia's dairy industry

This chapter gazes into future dairy farming with particular emphasis on the potential role of feedpad technology.

The main points in this chapter:

- Investments in feedpad technology must include improved effluent management.
- With the benefits of hindsight, most farmers should invest in larger feeding out machinery as their herds invariably increase over time.
- Feed efficiency should improve as feeding systems become more sophisticated with increasing capital outlay in feeding infrastructure and equipment, through increasing herd milk outputs, providing greater control in rumen metabolism and reducing feed wastage.
- The future of Australia's dairy farming will be in fewer, larger farms with increasing emphasis on flexibility in feeding systems.
- To make better use of the technology, farmers are seeking robust decision-making tools as well as more effective transfer of knowledge of feedpad systems from researcher to adviser to farmer.
- The changing climate variations in irrigation water availability requires further research into alternative feed sources, such as their production, access and integration into various production systems and also the feed conversion efficiencies of different production systems.
- This chapter lists the key research and development issues which highlight the importance of better understanding of milk responses in grazing cows to partial mixed rations.
- With more dairy farmers planning to leave the industry in future years, there are opportunities for some of these farmers to still remain part of the industry and service the feedpad technology without actually milking cows.

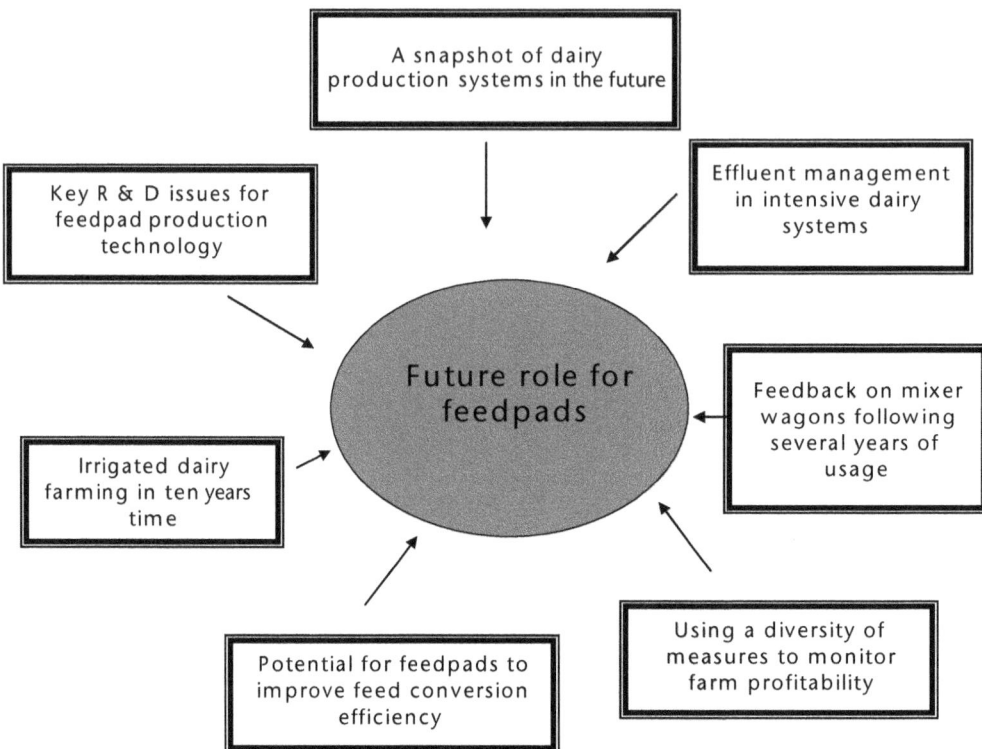

Figure 13.1 Key factors influencing the future role for feedpads in Australia's dairy industry.

The slowly recovering global economy hence low unit milk returns, the strong Australian dollar together with the current uncertainties of southern Australia's climate have put considerable pressure on Australia's dairy farmers during the 2009/2010 milking season. More farmers are likely to leave the industry than in previous years (Dairy Australia 2009b), particularly those approaching retirement age and with limited succession plans, and with high production costs in irrigated dairy regions. As is usual when the cost-price squeeze increases the need for greater farm efficiencies, the more innovative farmers look to increase cash flows via higher milk flows, either through greater per cow production or larger herd sizes. As such, farmers generally look to intensify their operations, and greater adoption of feedpad technology will be one such avenue.

Feedpads are now an integral part of Australia's dairy industry. Whether they will predominantly become the most sophisticated type, in which cows are zero grazed and fed entirely on TMRs, is unlikely, because of the relatively low costs of grazed pastures and the associated management skills developed by dairy farmers for managing such a feedbase. It is also highly unlikely that dairy feedlots will become the sole (or even predominant) dairy production system in any particular

Figure 13.2   Feedpads are now an accepted part of Australia's dairy industry.

dairy region of Australia, because land prices are relatively cheap and labour costs are expensive compared to other countries. These two major factors have led to zero grazing production systems throughout Asia and in other selected dairy regions such as California. The high requirements for machinery to produce conserved forages and their associated high purchase and maintenance costs, will also limit their role to being one of the smorgasbord of forages offered to Australian milking cows.

Because well-managed grazed pasture are likely to remain a relatively cheap forage source, provided input costs (such as for irrigation water or fertiliser) do not become prohibitive, future feedpad management will be based on integrating PMRs into current grazing systems. A combination of these feedstuffs (PMRs and grazed pasture) is more complex to manage compared to a single TMR. To date, little research has been undertaken to more fully understand the nutritional interactions inside the rumen when grazed pastures and TMRs are combined in a feeding system. Considerable knowledge has been gained about the combination of grazed pasture plus concentrates (cereal grains, agro industrial by-products with or without rumen additives); however, rumen metabolism is more stable when additional forages are included in the supplement. Milk responses would then be expected to be better with a PMR supplement rather than one based entirely on high-starch feeds.

This final chapter undertakes some 'crystal-ball gazing' into what is likely to happen when Australia's dairy industry enters the new phase of reduced natural resources due to climate change and other, as yet, not well-understood natural phenomenon, a more expensive feedbase and probably more intensive production systems. The total area devoted to dairy farming is likely to become smaller in Australia as urban pressures increase land prices and the natural resources restrict the more suitable dairying regions. Because global demands for food, particularly high quality food such as milk, cheese and other dairy foods, will increase, particularly as national incomes of previously underdeveloped countries improve, there will always be a demand for dairy products and Australia is well placed to provide such an important commodity. Therefore more intensive production systems in the most suitable environments to provide the high nutrient demands of milking cows, seem inevitable.

This chapter discusses a range of topics relevant to the future of feedpad production technology. These include concerns about effluent management as systems become more intensive; feedback from dairy farmers with a long history of feedpad usage; using feedpad technology to improve the conversion of feeds to milk; examining a scenario of irrigated dairy farming in 10-years' time; and lastly, some of the key research needs to make better use of existing and future feedpad production technology.

Major concerns from DPIV effluent management advisers

Anonymous (2005) reported on a survey of dairy farmers in northern Victoria regarding their proposed plans to incorporate feedpads into their dairy production systems. The farmers were asked the following questions:

1. Will the farm develop or upgrade its intensive feeding system within the next two years?
2. What type of feedpad was proposed; hay rings, formed earth/rock, concrete, roofed concrete?
3. Will the feedpad incorporate a hose or flood washing system?
4. Will the effluent be incorporated into the dairy shed system?
5. On average, will the cows be on the feeding area for more than 6 hr/d?
6. Was the storage and management of solids from the feeding area a concern?
7. Will the farm utilise more supplementary feeds than in the past?
8. Was the farmer aware of the SIRIC (2002) and O'Keefe et al. (2010) dairy cattle feedpad guidelines?

Of the farmers surveyed:

- 55% were expecting to develop or upgrade their feedpad system within the next two years.
- 30% had no feedpads and had no intention of developing one.
- 15% had no plans to upgrade their existing feedpads.

Managing nutrient runoff from earthen feedpads was a low priority. Of those constructing a pad, 20% planned to include a hose/flood washdown system while 42% would incorporate the feedpad effluent into their existing dairy shed effluent system. Of those with existing or proposed new feedpads, 22% would allow their herds more than 6 hr/d access to the pad while 54% would increase their imported supplements. In addition, 47% expressed concerns about managing feedpad effluent, while 33% were not aware even of the SIRIC (2002), let alone the O'Keefe *et al.* (2010), guidelines.

The DPIV adviser concerns could be summarised as follows. Many effluent systems are compromised because feedpads were not included in the original design. Management of feedpad systems requires a different approach to grazing only systems, such as feed storage, issues with animal behaviour and health and maintenance of facilities. Odour can easily become a problem thus involving government legislation. Lack of investment in early planning may be costly. Systems must remain flexible; for example, purchasing portable linear stock feeding modules limits options for developing the feeding system. Purchasing silage carts or mixer wagons is an integral part of feedpad development. Temporary feedpads can become permanent. The siting of feedpad must consider all eventualities, such as wet winters. Installing concrete surfaces changes effluent management. Money spent on upgrading effluent management is money well spent. Other feedpads should be visited to learn from other farmers' mistakes so they will not be repeated again. Farmers with operating feedpads should be asked specifically what they would change if they did it all again?

Mixer wagon survey of Queensland dairy farmers

A survey was undertaken of 10 dairy farmers in Queensland who had purchased a mixer wagon several years previously (Busby *et al.* 2007). This was effectively a survey of why these farmers wanted to intensify their dairy production systems, what they have achieved and if starting over again, how would they do it differently. Their reasons for intensifying their system were the same as those discussed in Chapter 3. The survey findings are summarised below.

Production responses:
- *Milk production.* All 10 farmers reported increases.
- *Milk fat content.* Nine noted higher contents, with five reporting increases of between 0.2 and 0.5%.
- *Milk protein content.* Four reported increases.
- *Reproduction.* It was too short-term to observe differences in conception rates.

Time to recover repayments:
- Six had no opinion.
- One had fully recovered them.

Figure 13.3 Plan ahead when purchasing feed out carts for mixer wagons.

- One noted an immediate production response.
- One took 3 months to achieve any production response.
- One saved $30/t because the wagon incorporated a roller mill.

All farmers said they would do it differently next time, for a variety of reasons. Irrigation was a more important investment, if feasible. Access to more by-products would make it more profitable. They would purchase a bigger machine, and they would have done more homework on the various types of machines available. They also said they should have sought out cheaper feeds.

Their advice to other farmers contemplating a mixer wagon were to resist the temptation to purchase a smaller machine as herds increase over time and some by-products are quite bulky; to remember that the wagon will not mix properly if it is more than two-thirds full. It is important to consider requirements for wet weather, such as surface slopes. Closer attention should also be given to details such as ration formulation and contract feed prices.

Tractor requirements:
- A four-wheel-drive tractor is necessary
- Most wagons mix adequately with a 60 hp tractor, but for safety, an 80 hp tractor is required
- A full bin weighs 7–10 t, so be aware of safety issues

- Round bales of hay or silage require more power to process than do square bales
- A 14 m³ capacity mixer wagon is required for herds with more than 100 cows.

Questions they want answered:
- What are the responses in milk production?
- What are the responses in milk fat and protein contents?
- Will it save feed?
- What are the benefits in herd health?
- What size mixer wagon should be used?
- What are the real costs?
- What investments are required in other facilities?
- How is long hay fed in a PMR?
- Are there any benefits in splitting up the milking herd? In other words, is there any advantage in feeding different ration formulations?

Advantages of mixer wagon-based systems include the fact that they provide better control with knowledge of weights and of nutritive analyses of individual feeds, and since milk production is linked to feed intake, increasing intakes has a positive effect. Slug feeding is eliminated and feeding preferences are reduced, leading to fewer digestive upsets. Ration flexibility, such as palatability, can be easily manipulated with molasses and some by-products. Labour becomes more specialised with benefits following on from greater attention to detail, and systems adapt well to computer ration formulation.

Financial considerations:
- Farmers must plan finances carefully, even when using the traditional principal and interest loans or chattel mortgages. The loans should be tailored to the effective life of the machine.
- They must also carefully consider the terms and conditions of any loan.
- They must assess their income stream realistically.
- They should consider the assets/equity position.

In summary, they concluded that:
- Milk yield per cow increased
- Milk protein was variable because milk yield will automatically reduce protein content
- Milk fat content generally increased, but not always.

The potential for feedpads to improve feed conversion efficiency

As feed costs can increase without associated increases in milk returns (such as during 2008/2009), there are only limited opportunities to improve farm profits.

Figure 13.4 Partial mixed rations are likely to result in more consistent milk responses in grazing cows than concentrates.

Initially, any extraneous variable costs (such as excess herd and shed costs) and fixed overhead costs need to be addressed. Then, the only real option available is to increase the amount of milk produced per unit of feed consumed, whether homegrown or purchased. Little (2009b) has outlined feed conversion targets as part of the Grains2Milk dairy research and extension program, funded by Dairy Australia. These targets vary with the type of feeding system imposed on the herd. These have been discussed in more detail in Chapter 2 and can be summarised as follows:

1. Low grain – where cows were offered grazed pastures plus other forages in the paddock together with up to 1 t concentrates/cow/yr fed in the milking parlour.
2. Moderate to high grain – where cows were offered grazed pastures plus other forages in the paddock together with more than 1 t concentrates/cow/yr fed in the milking parlour.
3. Partial mixed ration (PMR) – where cows grazed pastures for most of the year and were fed a PMR with or without additional concentrate feeding in the milking parlour.

4. Hybrid system – where cows grazed pastures for less than nine months each year and were fed forages and concentrates incorporated into a PMR.
5. Total mixed ration (TMR) – where cows were zero grazed hence continually housed and fed their forages and concentrates entirely as a TMR.

Feed conversion can be improved in many ways such as reducing feed wastage; maintaining feed quality; minimising feed gaps throughout the year; ensuring the amount of feed on offer does not limit intake; ensuring the ration is balanced for essential nutrients; keeping the rumen as stable and productive as possible; ensuring changes in body condition in early lactation do not exceed recommendations; minimising climatic stress; and optimising herd fertility to maximise the number of cows in early lactation at any one time.

The feed-conversion ratio is calculated as the annual milk production (kg) divided by the total feed offered to milkers. This includes the total sum of pasture, hay, silage, grain and other feeds all expressed as kg DM. This generally requires some estimate of the amount of pasture grazed throughout the year, which can be easily calculated using the Target 10 (2005) methodology. A new and more sophisticated approach to quantifying grazed pasture consumption has recently been published by Heard and Wales (2009).

For milking cows, the annual feed conversion includes a 60-day dry period. Milk yield can be expressed in kg milk or kg milk solids (that is milk fat plus milk protein).

Table 13.1 presents the annual targets for feed conversion for different feeding systems, both the achievable target and the critical feed conversion at which action should be taken to assess the factors restricting the value.

Feed efficiency is but one of the measures of feeding success and its sensitivity depends on the ability to estimate grazed pasture intake accurately, in the first four of the five systems in Table 13.1. The higher the proportion of grazed pasture intake of total feed intake, the greater the potential errors in the feed-efficiency estimates. Furthermore, as measures of supplementary feeding are invariably based on amounts offered rather than consumed, additional inaccuracies occur due to any unaccounted

Table 13.1 Annual feed conversion targets (kg milk/kg DM fed) for different feeding systems.

Feeding system	Achievable target	Take action if less than
1. Low grain	1.0	0.9
2. Moderate to high grain	1.2	1.1
3. Partial mixed ration	1.3	1.2
4. Hybrid system	1.4	1.3
5. Total mixed ration	1.6	1.45

(Source: Little 2009b)

wastages. Other potential errors could occur with unaccounted changes in pasture substitution rates and associative effects, although these should be reflected in reduced marginal milk responses, as discussed in Chapter 6. Little (2008) even cautioned its use for comparing farms with different feeding systems, concluding that feed efficiencies are best used within a given farm, from year to year.

Feed efficiency should improve as feeding systems become more sophisticated with increasing capital outlay in feeding infrastructure and equipment, through increasing herd milk outputs, providing greater control in rumen metabolism and reducing feed wastage. Accordingly, investment in feedpad technology should be expected to improve farm profits provided the marginal financial returns from the greater feed efficiency exceed the costs of such investments.

A scenario of irrigated dairy farming in 10 years' time

Poole and Cowan (2008), reporting on the future of Northern Victoria's dairy industry, presented the characteristics of successful dairy farmers in the region in 10 years' time, under three different water scenarios (namely a return to more favourable climatic conditions, medium climate change and continuation of recent low irrigation water allocations). Such characteristics are likely to be developed in northern Victoria, although the following list applies equally to dairy farmers Australia-wide. The common characteristics of successful dairy farmers, through all the three water scenarios above, were:

- They will operate fewer but larger farms.
- They will develop the ability to profitably implement flexible feeding systems, moving from pasture based to TMR depending on the resources available from one season to another.
- They will develop a more thorough knowledge of business cost structure, profit drivers and risk management strategies.
- There will be increasing reliance on farm automation, for their irrigation, dairy and feeding systems.
- They could operate sustainable dairy systems.
- They could manage a carbon-trading environment.
- They will develop stronger relationships with dryland fodder and grain producers.
- Their farm labour force will develop a higher skills base.
- They will be able to achieve improved productivity through advances in water-use efficiency, feed conversion efficiency and fodder species.

When prioritising research and development (R & D) needs for flexible feeding systems, Poole and Cowan (2008) concluded that in order for dairy businesses to be able to respond quickly to changing seasonal conditions, farmers will require an understanding of the impact such changes have on their business, what the cost structures will be for different systems, their associated risks and the infrastructure/capital required. Each business should then have to develop the capacity to make the necessary investments in capital to ensure they can be set up to be flexible and continue to grow.

In addition, the surveyed farmers requested access to the right tools to gain this information, such as decision-making tools for different farming systems, as well as clear and efficient transfer of knowledge from researcher to adviser to farmer.

The changing climate plus water availability requires further research into alternative feed sources. Such research should address their production, access and integration into various production systems and also the feed conversion efficiencies of different production systems.

Throughout Australia, irrigation technology needs to be better understood to improve efficiencies of application. Farmers require the tools to make the right decisions regarding its use under different irrigation regimes. They also need to be able to move from periods of no irrigation to irrigation. In addition, they should know more about the forage production responses to irrigation in different soil types.

Other priorities identified in resource use included improved use of energy on the farm and providing the best environment for cows in a hotter climate.

The primary needs for R & D in flexible feeding systems could be summarised as:

- improved business management including risk management strategies
- improved farm productivity
- the ability to implement annual changes in the mix of pastures, crops, forages, by-products and grain in response to varying availabilities and costs of irrigation water.

Modelling different dairy feeding systems

A more structured approach was undertaken by Beale *et al.* (2009) in their review of the future for dairy farming in the Murray Basin. They developed a model to quantify the profitability of dairy production (based on internal rates of return) of the five dairy feeding systems described by Little (2009a) but under three different business scenarios, ranging from the most pessimistic (declining real milk price and reduced availability of irrigation water), to less pessimistic (flat milk price and slightly reduced irrigation water availability) to most optimistic (wetter climate change scenario and steady growth in demand for dairy products).

The results of the model clearly showed that as the feeding system became more intensive (shifting along the spectrum from System 1 to 5), there is greater scope for better farm efficiency and for lower unit production costs. This then translates into higher returns and more secure cash flows in these intensive

Table 13.2 Key farm management principles necessary to benefit most from investing in feedpad technology.

Issue	Principles of profitable dairy production
Skills	More intensive operations require improved skills in pasture and forage crop agronomy.
	More intensive operations require better feeding management.
Intensiveness of operation	More intensive systems are not necessarily more profitable as they introduce need for higher precision hence more risk.
	The trade-off between homegrown and purchased feed tends to underestimate the cost of homegrown feed. More abundant homegrown feed is better for cash flow but the total feed costs must include land and water.
	Irrigated versus rain-fed systems can be more or less profitable because they require different asset investments and seasonal operations. For example, rain-fed forages generally require more land per cow.
Use of capital	Changes in investments can be achieved through better use of capital, leasing low yielding feed production land and increasing yields from land, cows and irrigation water.
	The relationship between yield (from cows, water and feed) and capital invested in the dairy enterprise is critical to ensure scope for long-term wealth creation.
Seasonal conditions	By itself, variation in seasonal conditions is not the major driver of fortunes. The comparative performance over time is more dependent on how well the assets are used, sourcing cheap feed and protecting cash flows.
	As seasonal conditions become more unreliable and as producers are more vulnerable to supplies of water and cheap feed, the more risk averse they become in risking assets and making incorrect decisions regarding herd and farm management.

(Source: Beale *et al.* 2009)

operations. To achieve greater returns, however, they require greater levels of skill in herd, feed, capital and people management. Some of these key management principles are summarised in Table 13.2.

Future R & D needs for feedpad production technology

The following lists some of the key research needs to make better use of existing and future feedpad production technology:

- Documentation of wastage rates leading to a better understanding of factors affecting wastage of forages and TMRs when fed out using under various feedpad scenarios. Only then will sufficient robust data become available to undertake the partial budgets required to develop guidelines for investing in feedpad technology. This will also provide valuable inputs to the Leddin and Armstrong (2009) computer model on investing in feedpad development.

- Monitoring a series of farming systems ranging from pasture with light feeding of grain and concentrate to a full TMR system.
- Additional systems research should incorporate a PMR system using a feedpad with the flexibility to change, depending on climatic conditions, from a high level of mixed ration feeding during drought years to a greater reliance on traditional grazed pasture under more normal seasons.
- Documenting the milk response in grazing dairy cows to specific by-products, in comparison to cereal grains. This includes pasture substitution as well as actual milk responses to supplementation, which would be influenced by associative effects as well as actual responses to supplement nutrient intakes. Milk responses include both milk yield and composition and immediate and delayed responses.
- The milk response in grazing dairy cows to PMRs, in comparison to high energy or high protein concentrates. Milk response covers includes pasture substitution, associate effects and milk solids responses to the nutrients supplied by PMRs.
- Developing nutrition programs for pasture-based dairy systems for efficient and profitable levels of milk production while maintaining herd health and reproductive performance. With dairy industries more focused on producing milk for manufacturing and export, rather than liquid markets, increasing emphasis should be placed on production of milk solids rather than just volume.
- Reviewing and if necessary, researching the financial benefits of recycling effluent from cows fed TMRs in different feedpad systems.
- Reviewing and if necessary, researching issues of animal welfare of grazing cows when confined to alleviate climatic stress or to provide a more comfortable environment than on pastures during prolonged wind and rainfall events.
- The concept of milk functionality is relatively new. This includes such factors as the yield of cheddar cheese from each litre of milk. With the current interest in developing new dairy products and markets, the ability to manipulate milk composition in a particular direction will become more relevant for farmers to be able to potentially value add more to their farm gate product.
- With the increasing likelihood of more milk producers planning to leave the dairy industry in future years, there are still opportunities for some of these farmers to still remain part of the industry. Instead of actually milking cows on their farm, they could redirect their efforts and resources to provide goods and services for other dairy farmers. Such examples include:
  - *Feed production.* Much of the forages and cereal grains sourced by dairy farmers have to be imported long distances from traditional grain growing

Figure 13.5 A sign of confidence in Australia's dairy industry – laying down the gravel for a new feedpad.

or mixed farming enterprises. It may be possible for some of these feeds to be produced nearby on what used to be milk-producing farms.

– *Contract rearing of replacement dairy heifers* is an integral part of Australia's dairy industry whereby dairy farmers contract other livestock producers to rear their replacement stock from weaning (at three months of age) until they are due to calve (at 24 months of age). Previous milk-producing farms could become specialist contract heifer rearing farms to service more of the dairy industry.

– *Growing out heifers for export.* Each year Australia exports 60 000 to 100 000 dairy heifers to many destinations around the world. Many of these stock are raised on existing dairy farms and have provided a valuable source of cash flow during periods of low milk prices. As well as contract rearing heifers for the domestic market, some farmers could purchase excess dairy heifer calves from milk producers to grow out for specific export markets.

– *Agisting dry cows.* An increasing number of previous milk producing farms could also service the industry by agisting mature cows during their two-month dry period each year.

– Provision of such goods and services would allow milk producing farms to concentrate their efforts and farm resources into what they do best, namely convert forages and concentrates into milk.

Much of the production technology for provision of these good and services is already known. What has yet to be fully researched is the business aspects of dairy farmers changing from just producing milk to servicing the entire industry. Farms with existing feedpads and associated machinery may be able to profitably redirect their dairy enterprise to growing out dairy heifers or even produce dairy beef from excess bull calves or cull dairy cows. If the feedpad investment has already been paid off, cash flows need not be as high as for farmers still servicing their debt of investing in the feedpad technology.

Abbreviations and conversions

Abbreviations

mm	millimetre
cm	centimetre
m	metre
ml	millilitre
ppm	parts per million
K	kilo or thousands
M	mega or millions
MCal	megacalories
MJ	megajoule
MT	megatonnes
min	minute
hr	hour
d	day
yr	year
mg	milligram
kg	kilogram
g	gram
J	joules
L	litre
lb	pound
hp	horsepower
ft	foot
hd	head
$	dollar
c	cent
<	less than
>	greater than

Conversion of Imperial units to metric units

Length:	1 inch = 25.4 mm
	1 foot = 30.5 cm
	1 yard = 0.91 m
	1 mile = 1.61 km
Volume:	1 cu ft = 0.028 cu m
	1 pint = 0.57 L
	1 gallon = 4.54 L
	1 bushell = 36.4 L
	1 acre foot = 1.23 Ml (megalitre)
Area:	1 acre = 0.40 ha
	1 sq mile = 2.59 sq km
Weight:	1 ounce = 28.3 g
	1 pound = 0.454 kg
	1 hundred weight = 50.8 kg
	1 long ton = 1017 kg (2240 lb)
Energy:	1 calorie = 4.19 joules
Density:	$1\ lb/ft^3 = 0.063\ kg/m^3$
Rate:	1 gallon/acre = 11.23 l/ha
	1 pound/acre = 1.12 kg/ha
	1 gallon/ton = 4.17 l/tonne
Pressure:	1 pound/sq in (psi) = 1.45 kPa (kilopascals)
Yield:	1 lb/ac = 1.12 kg/ha
Temperature:	$1°F = ((9/5) * C) + 32$
	1 F° = 0.56 C°
	50°F = 10.0°C
	60°F = 15.6°C
	70°F = 21.1°C
	80°F = 26.7°C
	90°F = 32.2°C
	100°F = 37.8°C
	110°F = 43.3°C

Conversion of US units to metric units

Volume:	1 gallon = 3.79 L
	1 bushell = 35.2 L
Weight:	1 hundred weight = 45.4 kg
	1 short ton = 907 kg (2000 lb)
Milk prices:	$10/hundred weight = 22.0 c/L

Glossary

Terms in this Glossary are defined in the context of their use in this book.

acidosis. An excessive increase in rumen acid caused by feeding too much grain or other starchy feeds or by introducing them into the diet too quickly.

ad lib or *ad libitum*. Fed to appetite.

acid detergent fibre (ADF). The less digestible or indigestible parts of the fibre; i.e. the cellulose and lignin only.

anaerobic pond. A lined pond in which anaerobic conditions prevail (minimal or no oxygen). These ponds are usually deep to minimise surface area, which reduces temperature fluctuation and enhances anaerobic decomposition.

animal welfare. The welfare of an individual animal with regard to its attempts to cope with its environment.

associative effect. Reduction in digestibility of one feed due to increasing intake of another type of feed.

average milk response. This is calculated from the quantity of extra milk produced by total supplement intake.

Best Management Practice (BMP). Management practices which reflect the best available knowledge, information and technologies.

body condition. Energy stored in body reserves by cows, predominantly as fat.

Biological Oxygen Demand (BOD). An index of the oxygen demanding properties of biodegradable material in water.

brisket board. The board in front of the free stall to prevent stock from lying too far inside the stall.

buffer distance. The recommended distance between a feedpad complex and existing housing, land zoned for residential or urban purposes, or other sensitive uses.

calving pad. An area on the farm specifically for calving down the cows, to provide a warmer, drier option to the paddock to facilitate round-the-clock access for the care of new born and young calves.

comfort zone. Range of air temperature when there is no measurable fluctuation in physiological processes of cattle.

condition score. Objective visual assessment of a cow's body condition on a scale of 1 (emaciated) to 8 (obese).

contingency plan. A plan for procedures and contact details for unexpected events, such as machinery breakdowns and disease outbreaks.

controlled drainage area. A self-contained catchment surrounding drainage areas of a feedpad complex. It is typically established using a series of catch drains and/or diversion banks or drains.

cow alley. An alleyway which provides cows with access to free stalls.

cow barrier. The structure to prevent stock standing in their feed, which also contains the stock in the feedpad.

crude fibre (CF). A measure of fibre in the diet now considered unacceptable as it does not always take into account all of the constituents that make up the fibre component of a feed; it measures only the alkali-soluble lignin and the cellulose.

crude protein (CP). A rough measure of all the protein in the diet; it assumes (incorrectly) that all the nitrogen in a feed comes from protein.

cubicle (or stall). The resting area for stock in free stalls where they can enter and leave at will.

dairy cattle unit (DCU). A measure of dairy stock based on their average live weight and the duration of each day they spend on a feedpad.

desludge. Removal of solid or slurry effluent from a pond, sump, drain or solids trap.

development. This includes construction of a building, carrying out works, subdividing land or buildings or displaying signs.

digestibility. The proportion of the dry matter in a feed that gets digested; it is the difference between what is eaten and what comes out as manure.

DPIV. Department of Primary Industries Victoria.

drive alley. The area adjacent to the feeding area where machinery delivers the feed, which may or may not include the feed alley.

dry matter (DM). The proportion of any feed remaining after all the water has been taken out.

effluent. The faecal and urinary excretion of livestock. This may also contain bedding, spilled feed, water and soil. It may also contain other components not associated with livestock excreta, such as dairy shed wash-down water, contaminated milk or hair.

Effluent Management Plan. The technical design and management of effluent on the farm, focusing on effective use of nutrients.

energy. The part of a feed that is used as 'fuel' in carrying out the cow's bodily functions.

Environmental Management Plan. A plan that focuses on general management of the farm, taking into account the environment and associated risks.

feed alley. The alley occupied by the stock to access the feed, usually located parallel to the feeding area.

feedpad. That part of a dairy farm used for supplementary feeding of stock on an area of land that is formed, surfaced or stocked at a rate to preclude vegetation. It is used for the purposes of dairy production and/or growth of young stock and for the protection from adverse environmental impacts, such as wet, cold or hot conditions. This is also known as a standoff or wintering pad.

feedpad complex. This is the entire feedpad which includes the feeding areas (such as drive and feed alleys, actual feeding area and watering points), internal alleys, loafing/bedding areas, feed storage and preparation areas, drains and ponds and manure stockpile. It does not include the manure and recycled effluent utilisation areas.

feeding area (table or strip). The place where the feed is placed in front of the stock. It can be on bare ground, on rubber matting, in troughs or on concrete surfaces.

feeding profit. A term used to describe the calculation of milk income less purchased feed costs.

feedlot. Land on which cattle are restrained by pens or enclosures for the purpose of intensive feeding. This does not include any area where stock are penned or enclosed for grazing, hand-feeding prior to 12 weeks of age for weaning, or the provision of subsistence rations due to fodder shortages, abnormal seasonal conditions or like events or the provision of supplementary rations for cattle which have daily access to pasture.

free stall cubicles. Individual cow bedding areas where partitions orientate stock for comfort and sanitation, providing each cow with a dry and comfortable place to lie down and rest and ruminate.

free stall sheds (sometimes called barns). Partially or fully enclosed sheds in which cattle are housed and provided with feed and water. They include the actual free stall cubicles, other loafing areas, drive and feed alleys, actual feeding area and watering points as well as the internal alleys.

fibre. The cell wall, or structural material, in a plant made up of (among other things) cellulose, hemicellulose, and lignin.

genetically modified (GM). A genetic engineering technique used to alter DNA by transferring genetic material from one organism into another in a laboratory.

geosynthetics. Thin, flexible and permeable sheets of synthetic material used to stabilise and improve the performance of soil in civil engineering works.

groundwater. All water below the land surface that is free to move under the influence of gravity.

guidelines. Written documents that tend to concentrate more on statutory planning, siting and engineering design, animal health and welfare and farm safety issues with less emphasis on feeding and other aspects of dairy production technology.

intensive animal husbandry. The practice of keeping or breeding farm animals, by importing most of the feed from outside the enclosures. It does not include abattoir or sale yards, emergency or supplementary feeding or the housing of animals for weaning, dipping or other husbandry purposes, if this is incidental to the use of the land for extensive animal husbandry.

LWT. Live weight.

leaching. The process where soluble nutrients are carried by water down through the soil profile.

loafing pad. A formed hard surface adjacent to the feedpad complex where stock can ruminate, seek shade and if contained for lengthy periods, lie down.

longitudinal slope. The slope along the length of the feedpad.

loose housing. Loafing pads where stock are free to lie down anywhere on the pad, in contrast to free stalls where cows are allocated specific areas to lie down.

Lower Critical Temperature (LCT). The air temperature at which the energy intake must increase to minimise reduction in weight loss in growing cattle or to prevent weight loss in mature cattle.

manure. See *effluent*.

marginal milk response. The extra milk produced in response to the next kg of supplement fed.

mastitis. Inflammation of the mammary gland.

Maximum Residue Limits (MRL). Set for a range of chemicals in food commodities and animal feeds that must not be exceeded in food products such as milk and meat.

metabolisable energy (ME). The amount of energy provided by a feed after deducting energy lost to faeces, urine, heat, and gas production; it is the energy available to be used by the cow for her metabolic activities.

MJ ME/kg DM. The standard measurement of ME in megajoules of metabolisable energy per kilogram of dry matter.

milk solids (MS). The amount of milk fat and milk protein supplied in raw milk, usually measured in kg /cow/day

N. Nitrogen.

neck rail. The rail in a free stall to assist positioning the cows so they have sufficient forward lunging space when they lie down.

neutral detergent fibre (NDF). A measure of all the fibre (hemicellulose, lignin, and cellulose) in a feed; it indicates how bulky the feed is.

nib wall. A small concrete wall constructed along the perimeter of alleys to prevent manure entering the feeding area and containing it within the feedpad area.

non-protein nitrogen (NPN). Not actually protein but simple nitrogen; however, microbes can make protein from simple nitrogen if enough energy (carbohydrates) is available in the rumen at the same time. Urea is a form of NPN.

Nutrient Management Plan. A plan for nutrient budgeting and mapping, focusing on the productive use of nutrients across the farm.

nutrients. the key components of stock feeds used for productive purposes. Also used to describe the elements (nitrogen, phosphorus and potassium) in effluent that can be useful to plant growth if applied as part of a planned approach, but are harmful to water quality.

O H & S. Occupational Health and Safety.

partial budget. A budget drawn up to estimate the effect on profit of a proposed change affecting only part of the farm. It is used to estimate the extra return on extra capital invested.

partial mixed rations (PMR). The formulated ration supplied to grazing stock while on the feedpad.

persistency. The monthly rate of decline of milk yield from peak values, in early lactation.

pH. A measure of acidity or alkalinity on a scale from 1 (extremely acid) to 14 (extremely alkaline).

pollution. The introduction of a pollutant into the environment that becomes harmful to the health of humans, other living organisms and the general environment. A pollutant may be chemical, physical, biological or energy (in the form of noise, heat or light).

protein. The material that makes up most of the cows body (muscles, skin, organs, blood); it also is part of milk.

quality. In relation to feeds, it is an indication of the level of energy and digestibility. In relation to milk, it refers to the level of various contaminants in milk, such as bacterial, chemical or any other adulterations that can be detected.

R, D & E. Research, development and extension.

recycled effluent. Liquid manure from a holding pond that is used to clean dairy yards, feed pad alleys or is applied to pastures or crops.

reuse. The application of manure or recycled effluent onto crops, pastures or other vegetation. The application rate is based on calculated nutrient budgets for that specific crop and soil type.

rumen degradable protein (RDP). The portion of protein in the diet that is digested and used by the microbes in the rumen to build themselves, if enough energy (carbohydrates) is available at the same time.

rumen undegradable protein. The portion of the protein that escapes rumen degredation.

runoff. All the surface water flow, both over the ground surface as overflow and in streams as channel flow. It may originate from excess rainfall that cannot infiltrate the soil or as the outflow of groundwater along the lines where the water table intersects the earth's surface.

sedimentation pond. A lined pond that is usually shallow and narrow to enhance manure drying and removal.

selection differential. This quantifies the ability of grazing stock to select the better quality pasture than that on offer.

sensitive use. A use that involves the presence of people, causing it to be sensitive to amenity considerations, such as odour, dust and noise. This includes a dwelling, a dependent person's unit, such as a residential building, hospital, school, day-care centre, caravan park or other uses involving the presence of people for an extended period. It does not include recreational areas such as parks and sporting facilities.

sensitivity analyses. Checking the effects on a planned outcome of a change in one or more of the factors that affects that outcome. Frequently used in partial budgets.

separation distance. The distance separating a possible source of an emission (such as dust, odour or noise) from a potential receptor. This is a critical factor as to whether that emission causes a nuisance or not. This is now an accepted means of mitigating the impacts associated with odour, dust or noise emissions. This is measured from the nearest physical part of the feedpad complex, not including the manure and recycled effluent reuse area.

shoulder rail. A rail in the free stall to encourage the animal to move backwards as she stands, hence defaecate and urinate in the cow alley not in the stall.

side slope. The slope in the feed alley that directs manure and runoff away from the feeding area, hence is perpendicular to the feeding area.

silage feeding face. Maximum area of open silage stack to minimise deterioration of the silage prior to its removal.

slurry. Effluent in the form that is too thick to pump or spray and is unable to be handled as solid effluent. May be able to be pumped with specialist equipment.

SOP. Standard Operating Procedures.

solids separation. Separation of solids and liquid components of effluent so they can be managed or utilised separately.

storage pond. A lined pond constructed to store runoff from a feedpad complex, which may or may not contain semi-solid or liquid manure. The contents are recycled for flood washing, or stored for application to land. Storage ponds are often connected in series with anaerobic or sedimentation ponds.

stocking density. A measure of the intensity with which a feedpad is stocked. It is normally expressed in terms of the area provided for stock as m^2/DCU. It is calculated using the areas provided on the pad or loafing area, where stock have access to feeding, watering or loafing facilities; that is, where stock might be held on a permanent basis.

stock containment area. An area on the farm set aside to assist with stock management during adverse climatic conditions, prolonged drought and in times of emergencies such as fire or flood.

stockpile. An area where solid effluent, feed wastage or bedding is stored before being spread on farm pastures.

substitution rate. This is a measure of the reduction in pasture DM intake per kg DM of the supplement offered.

supplement. A feed or product added to the cow's diet to increase the intake of some dietary component, such as energy, protein, fibre, vitamins or minerals.

surface waters. These include dams, impoundments, rivers, creeks and waterways where rainfall is likely to collect. It excludes groundwater and waters within tanks, artificial waste treatment systems, reticulated supply distribution systems, off-stream private dams and piped and underground drains.

Temperature Humidity Index (THI). A system for quantifying heat stress based on temperature and humidity. The higher the index, the greater the discomfort, and this occurs at lower temperatures for higher humidities.

total mixed ration (TMR). The fully formulated ration fed to stock in a feedpad, with no access to grazed pasture.

TS. Total solids content of feedpad effluent.

undegradable dietary protein (UDP). Any protein in the diet that passes through the rumen without breaking down and is digested in the abomasum and small intestine. Also known as bypass protein.

Vendor Declaration Form (VDF). This is a declaration allowing suppliers of stock feeds to declare valuable information on the safety of that feed for stock.

wastage. The loss of feed between its purchase and when is goes 'down the cow's throat', which takes into account losses during storage, mixing, delivery to the feedpad and on the feeding area prior to being consumed by the stock. This is usually expressed as DM rather than specific feed nutrient.

watercourse. A naturally occurring drainage channel with a clearly defined bed and bank, in which water can flow at any time.

waterway. This is defined as a river, creek, stream or watercourse, a natural (or modified) channel, lake, lagoon, swamp or marsh where water regularly flows. It can also include collections of water officially designated as waterways or land on which, as a result of works constructed, water can collect.

Whole Farm Plan. A plan that focuses on the overall management and layout of the entire farm including improvements and infrastructure.

withholding periods (WHP). Minimum length of time that dairy cow cannot be milked after consuming a particular feed.

References and further reading

Amaral-Phillips DM, Bicudo JR and Turner LW (2002) 'Feeding your dairy cows a total mixed ration: getting started'. Bulletin ID-141A. Cooperative Extension Service, College of Agriculture, University of Kentucky, Lexington, US.

Anonymous (2005) 'Practice change evaluation 2005. Goulburn Broken nutrient and water quality'. Project WQ007. Goulburn Broken Catchment Management Authority, Victorian Department of Primary Industries, Kyabram.

Anonymous (2006) 'Dairying for tomorrow. Survey of NRM practices on dairy farms'. Report commissioned by Dairy Australia.

Ashton D and Mackinnon D (2008) 'Australia dairy industry. Use of technology and management practices on dairy farms'. Research report 08.12. Australian Bureau of Agricultural and Resource Economics, Canberra.

Beale R, Radcliffe J and Ryan P (2009) 'A dry argument: future for dairy in the Murray Basin?' Report to Dairy Australia.

Bolsen K and Pollard G (2004) Feed bunk management to maximise feed intake. *Dairy Technology* **16**, 227.

Busby G, Barber D, Warren R and Walker R (2007) 'Mixer wagons – some observations'. M5 Info series 121. Queensland Department of Primary Industries, Brisbane.

Chamberlain P (2002) 'Farm animal management and welfare guidelines'. Queensland Dairyfarmers' Organisation.

Dairy Australia (2007a) 'Feed. Fibre. Future'. A series of Technical Fact Sheets for Grains2Milk program. Dairy Australia, Melbourne. http://www.dairyaustralia. com.au/Farm/Feeding-Cows/Feeding-Systems/Fodder-Shortage-Fact-Sheets. aspx

Dairy Australia (2007b) 'Flexible feeding systems'. A series of Technical Fact Sheets for Grains2Milk program. Dairy Australia, Melbourne. http://www. dairyaustralia.com.au/Farm/Feeding-Cows/Feeding-Systems/Flexible-Feeding-Systems.aspx

Dairy Australia (2008a) 'Buying feed. An information pack for dairy farmers about buying feed'. A series of Technical Fact Sheets for Grains2Milk program. Dairy

Australia, Melbourne. http://www.dairyaustralia.com.au/Farm/Feeding-Cows/ Purchased-Feed/Feed-Purchasing.aspx

Dairy Australia (2008b) *Effluent and Manure Management Database for the Australian Dairy Industry*. Dairy Australia, Melbourne. http://www. dairyingfortomorrow.com/index.php?id=48

Dairy Australia (2008c) 'Cool cows. Dealing with heat stress in Australian dairy herds'. Dairy Australia, Melbourne. http://www.coolcows.com.au/

Dairy Australia (2009a) 'Dairy 2009. Situation and outlook'. June. Dairy Australia, Melbourne.

Dairy Australia (2009b) 'Dairy 2009. Situation and outlook'. October. Dairy Australia, Melbourne.

Dairy Australia (2009c) 'Insights into the Australian dairy feed market, 2009'. Dairy Australia, Melbourne.

Dairy Australia (2009d) 'A review of 11 applied dairy nutrition models used in Australia'. Dairy Australia, Melbourne.

Dairy Gains (2008) 'Management of dairy effluent. 2008 Dairy Gains Victoria guidelines'. Victorian Department of Primary Industries, Kyabram.

Davison T and Andrews J (1997) 'Feedpads Down Under'. Queensland Department of Primary Industries, Brisbane.

DEXCEL (2005) 'Minimising muck, maximising money'. DEXCEL, New Zealand.

Doyle P, Stockdale R, Lawson A and Cohen D (2000) 'Pastures for dairy production in Victoria'. 2nd edn. Victorian Department of Primary Industries, Kyabram.

DPIV (2010) 'Guidelines for Victorian dairy feedpads and freestalls'. Department of Primary Industries Victoria, Echuca.

Gibb I (2009) 'Farming systems and risk'. Presented at dairy farmer workshops in northern Victoria on sowing options for autumn, February 2009.

Gibb I, Metcalf L, Shannon P and Moran J (2006) 'Marginal thinking: the path to more profitable dairy farming'. Discussion paper for Murray Dairy Farm Business Management project.

Griffiths N, Mickan F and Kaiser A (2004) Crops and by products for silage. In *Successful Silage*. (Eds A Kaiser, J Piltz, H Burns and N Griffiths) pp. 109–141. Dairy Australia and NSW Department of Primary Industries.

Heard J, Cohen D, Doyle P, Wales W and Stockdale R (2004) Diet Check, a tactical decision support tool for feeding decisions with grazing dairy cows. *Animal Feed Science and Technology* **112**, 177–194.

Heard J and Wales W (2009) 'Pasture consumption calculator'. Victorian Department of Primary Industries. http://www.dpi.vic.gov.au/ pastureconsumptioncalculator

Jenkin M and Wales W (2007) 'Diet Check instruction manual. Supplying metabolisable energy, crude protein and neutral detergent fibre to meet your milk production targets through the effective use of pastures and supplements'.

Victorian Department of Primary Industries, Kyabram. http://www.dpi.vic.gov.au/DPI/nrenfa.nsf/9e58661e880ba9e44a256c640023eb2e/28c97d65e3b0fe11ca257423007fd8f1/$FILE/Diet%20Check%20manual%20.pdf

Kaiser A, Piltz J, Burns H and Griffiths N (2004) *Successful Silage*. TopFodder. Dairy Australia and NSW Department of Primary Industries.

Leddin C and Armstrong D (2009) 'Feedout checkout: a decision support tool to assist in making investment decisions on feed pads and feeding machinery'. Victorian Department of Primary Industries. http://www.dpi.vic.gov.au/feedoutcheckout

Little S (2007) Establishing flexible feeding systems. *Australian DairyFarmer* **Nov/Dec**, 20–23.

Little S (2008) A new industry-wide approach to securing greater returns from grain and concentrate feeding. *Proceedings of the Australian Dairy Conference, Tasmania, Feb 2008*.

Little S (2009a) Feeding the right feeding system. *Australian DairyFarmer* **May/Jun**, 69–70.

Little S (2009b) Improving feed conversion. *Australian DairyFarmer* **Jul/Aug**, 42.

McDonald S, Spry J and Janssen M (2008) 'Developing and managing dairy cattle feedpads'. Technical Information Notes. Victorian Department of Primary Industries, Kyabram.

Mickan F (2009) Guts of alternative feeds. *Australian DairyFarmer* **May/Jun**, 77–81.

Moran J (1989) 'Feeding and calving pads for cattle'. DARA Agnote No. 4112/89.

Moran J (1996) *Forage Conservation. Making Quality Silage and Hay in Australia*. AgMedia, Melbourne.

Moran J (2005) *Tropical Dairy Farming: Feeding Management for Small Holder Dairy Farmers in the Humid Tropics*. Landlinks Press, Collingwood.

Moran J (2006) 'Feeding and calving pads for cattle'. DPI Agricultural Notes AG0015.

Moran J (2009) *Business Management for Tropical Dairy Farmers*. Landlinks Press, Collingwood.

Nutrient Requirements of Domesticated Ruminants (2007) CSIRO Publishing, Collingwood.

O'Keefe M, Chamberlain P, Chaplin S, Davison T, Green J and Tucker R (2010) *Guidelines for Victorian Dairy Feedpads and Freestalls*. Department of Primary Industries, Victoria.

Poole D and Cowan T (2008) 'Resources and structures needed to effectively deal with the R, D & E priorities of the northern Victorian dairy industry'. Dairy Australia & Victorian Department of Primary Industries.

SIRIC (2002) 'Dairy cattle feedpad guidelines for the Goulburn Broken Catchment'. Shepparton Irrigation Region Implementation Committee, May 2002. http://

www.gbcma.vic.gov.au/downloads/DairyFeedPadGuidelinesMay2002/FrontSection.pdf

Sprecher D, Holstetler D and Kaneene J (1997) A lameness scoring system that uses posture and gait to predict dairy cattle reproductive performance. *Theriogenology* **47**, 1178–1187.

Stevens D and Platfoot G (2005) Feeding out losses of whole crop cereal silage. *Proceedings of the New Zealand Grassland Association* **67**, 137–140.

Stockdale CR (2010) Wastage of conserved fodder when feeding livestock: a review. *Animal Production Science* **50**, 400–404.

Stockdale R, Dellow D, Grainger C, Dalley D and Moate P (1997) 'Supplements for dairy production in Victoria'. Victorian Department of Primary Industries, Kyabram.

Target 10 (2005) *Feeding Dairy Cows: A Manual for Use in the Target 10 Nutrition Program*. 4th edn. (Eds J Jacobs and A Hargreaves). Department of Primary Industries, Melbourne.

Index

www.ingramcontent.com/pod-product-compliance
Lightning Source LLC
Chambersburg PA
CBHW052141170526

45159CB00017B/3130